Introduction to the Physical Metallurgy of Welding

Butterworths Monographs in Materials

The intention is to publish a series of definitive monographs written by internationally recognized authorities in subjects at the interface of research interests of the academic materials scientist and the industrial materials engineer.

Already published

Amorphous metallic alloys
Control and analysis in iron and steel making
Die casting metallurgy
Metal resources and energy

Forthcoming titles

Continuous casting of aluminium
Energy dispersive X-ray analysis of materials
Eutectic solidification processing
Mechanical properties of ceramics
Metallurgy of high speed steels
Microorganisms and metal recovery
Residual stresses in metals

Butterworths Monographs in Metals

Introduction to the Physical Metallurgy of Welding

Kenneth Easterling, Dr.Tech.Sci
Professor and Head of the Department of Engineering Materials,
University of Luleå, Sweden

Butterworths
London Boston Durban Singapore Sydney Toronto Wellington

First published, 1983

British Library Cataloguing in Publication Data

Easterling, Kenneth E.
Introduction to the physical metallurgy of welding.
—(Butterworths Monographs in Materials)
1. Steel—Welding—Handbooks, Manuals, etc.
I. Title
672.5'2 TS227.A1

ISBN 0–408–01351–6
ISBN 0–408–01352–4 Pbk

Typeset by Gatehouse Wood Ltd., Vestry Estate, Sevenoaks.

Printed and bound by Page Bros (Norwich) Ltd.

Preface

I have written this book for undergraduate or postgraduate courses in departments of metallurgy, materials science or engineering materials. However, I hope that it could also serve as a useful revision text for engineers concerned with welding problems in industry. The book deals primarily with the welding of steels, which reflects the much larger volume of literature on this material, but many of the principles discussed can also be applied to other alloys. Indeed, the theme of the book is to deliberately avoid too much detail in order to grasp the basic physical metallurgical principles involved. Welding is such a complex process that I could see no other way to treat the subject.

The book is divided into four chapters, of which the middle two deal with the microstructure and properties of the welded joint, i.e. the weld metal and the heat-affected zone. The first chapter is designed to provide a wider introduction to the many process variables of fusion welding, particularly those that may influence microstructure and properties, while the final chapter is concerned with cracking and fracture in welds. The book ends with a comprehensive case study of the *Alexander Kielland* North Sea accommodation platform disaster.

In spite of the great importance of welding in the construction industry, until now there has been no student text that deals specifically with the *physical metallurgy of welding,* although there have been conference proceedings on the subject and it has been treated in a number of more general books on welding metallurgy. However, because of its importance and complexity the subject is an exciting and challenging one, and now merits inclusion in undergraduate and/or postgraduate curricula.

I am indebted to Dr. F.B. Pickering (Sheffield Polytechnic) for critically reading the original manuscript, and for making a number of valuable suggestions, and to John Ion (University of Luleå) for helping with the proof-reading.

In collecting material over the past few years I have had many useful discussions with a number of people and I should especially like to mention Carl-Henrik Rosendahl and Göran Almqvist of ESAB; Nils Leide of Kockums AB; Nils Christensen of the Norwegian Institute of Technology, Trondheim; Hans Åström of Carbidkontoret, Gothenburg; Nils-Erik Hannerz of Swedish Steel; and Michael Ashby of Cambridge University. I should particularly like to thank Dr H.S. Wintermark, Norske Veritas, Oslo, for discussions concerning the *Alexander*

Kielland investigation, and Dr L-E. Svensson, ESAB (Gothenburg), for sending
a report of this investigation as well as for giving me valuable advice on weldi
consumables development. In addition, I am grateful to Bengt Utterberg of ESA
for sending me a number of beautiful micrographs. I have also received considera
help from *The Welding Institute,* Cambridge, where I have often had use of th
excellent library facilities. Above all, however, I should like to thank my colleag
at the University of Luleå, particularly Bengt Loberg, Erik Navara, Jan Strid a
John Ion, for their support and many stimulating discussions. Last, but not leas
am grateful to Agneta Engfors for her skilful secretarial assistance and patient typi
of the manuscript.

Kenneth Easterling
Luleå

Contents

Prologue

400 AIRLIFTED FROM OIL-RIG

More than 400 construction workers were airlifted to Aberdeen from a North Sea accommodation rig in Shell's Fulmar oilfield yesterday after an 11-inch crack was discovered in a weld on one of its leg-braces. The rig, the *Borland Dolphin,* is to be towed to either Norway or Scotland for repairs. Shell said the move was not an evacuation but a 'demanning'; there was no question of any danger because the flaw had been spotted in time.

The fault in the *Borland Dolphin* hired by Shell from a Norwegian company, strongly resembles the kind of structural flaw that led to the *Alexander Kielland* disaster in March 1980 in which 123 British and Norwegian oil workers died.

The Sunday Times,
London, 6th December, 1981

Chapter 1

Fusion welding – process variables

The present scale of modern, welded construction is really quite staggering. The new 6000 km pipelines used to transport natural gas from the other side of the Ural Mountains to Western Europe, the giant 100000 ton supertankers, the oil platforms, and the aluminium liquid-gas storage tanks and rolling stock are just a few of the more impressive examples. The wide range of materials used to build these huge constructions have all to meet rigid specifications of high strength, good toughness and often resistance against corrosion and fatigue. Most importantly, however, it is essential that the materials possess *good weldability*.

What is really meant by 'good weldability'? There have been numerous discussions among the various welding commissions as to its exact meaning. Some people even argue that since virtually all metals and alloys can be welded, the term may even be redundant! However, this point of view is rather academic since, in practice, materials are generally considered to possess good weldability only if they can be reliably welded on a production scale. As such, the term *good weldability* has to be a function of a number of interacting factors, which include:

1. Type of welding process.
2. Environment.
3. Alloy composition.
4. Joint design and size.

All these factors can be decisive to the alloy's weldability and if one of them is unsuitable it may give rise to cracking problems. Indeed, weldability is often defined simply in terms of susceptibility to the various types of cracking problems known to be associated with welds. Thus, the use of empirical equations which describe an alloy's susceptibility to a certain type of cracking problem in terms of its composition, and in more complicated cases, in terms of a number of interacting parameters, such as carbon equivalent, heat input, preheat temperature, joint type, etc, is often advocated; this approach is examined in more detail later. In this chapter weldability is considered in terms of the various process variables of fusion welding and the weld thermal cycle.

Fusion welding

By far the most important technique used in welding construction today is fusion welding, which is the main process discussed herein. For descriptions of these and other welding processes *see*, e.g., Lancaster[1] and Tylecote[2] for solid-phase welding and Crossland[3] for a review of explosive welding. It is not intended to delve deeply into the metallurgy of fusion welding, but instead to discuss some of the factors that may affect the microstructure and properties of welds.

Figure 1.1 illustrates one of the most common of the fusion welding processes: *manual metal arc* welding (MMA). The heat source is provided by an *electric arc*, a high current and a low voltage discharge in the range of 10—2000 A and 10—50 V. The *electrode* consists of a core of filler wire and a

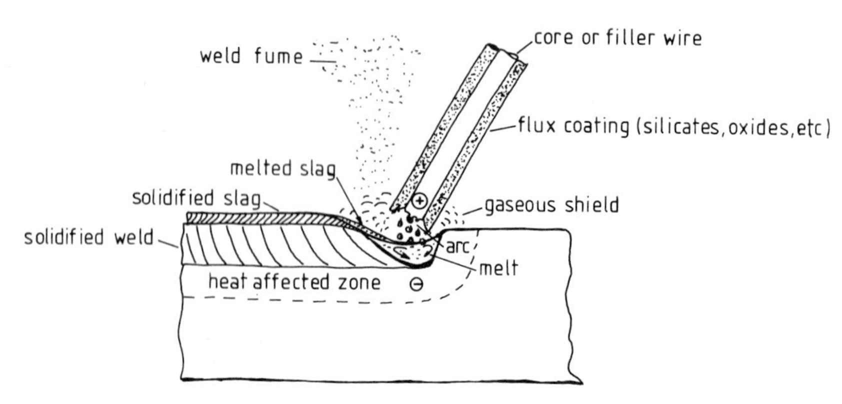

Figure 1.1 Principles of the manual metal arc welding process. Melting and vaporization of the flux coating provide a protective shield from the atmosphere

flux coating, composed of various silicates and metal oxides. During welding the flux melts to form a viscous *slag*, which provides a protective layer between the atmosphere and the molten metal. In addition, the slag creates a chemically reducing environment which helps to exclude elements in the air and to prevent moisture from penetrating the weld pool. It may also generate gases which help maintain the flow of liquid metal droplets from the filler wire to the pool. The slag has to adequately cover the melt, even in vertical welds, and so its composition is critical as this affects the viscosity. Afterwards, the slag must also be easily detachable from the solidified weld surface.

Other welding processes, such as *submerged arc* (SA), employ a separate slag feeding process (*see Table 1.1*), but the principle is the same as for MMA welding. A protective atmosphere may also be generated in the form of an inert gas as, for example, in *metal–inert gas* (MIG) welding. Alternatively, welding may be carried out in a vacuum as in *electron beam* welding. Some of the more common techniques of fusion welding are shown in *Table 1.1* with comments about the areas of application and possible problems.

It is important to note the different roles of the slag when comparing *arc welding* with *electroslag welding* (*see Table 1.1*). In electroslag welding the slag

TABLE 1.1 Principal fusion welding processes

	Manual metal arc	*Gas metal arc*	*Submerged arc*	*Electroslag*
	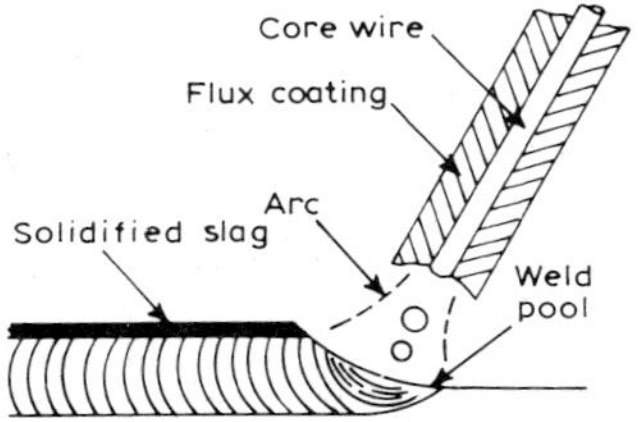 Core wire; Flux coating; Arc; Solidified slag; Weld pool	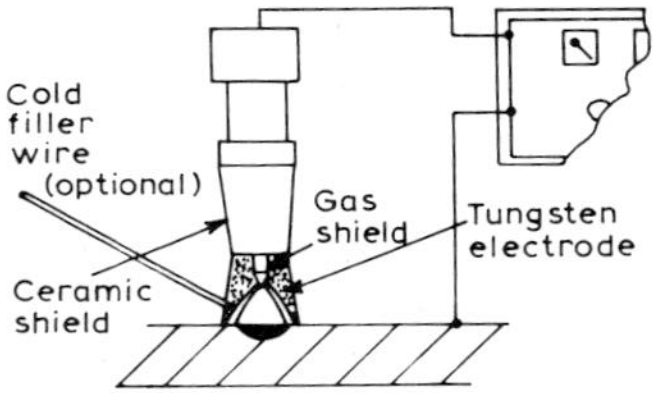 Cold filler wire (optional); Ceramic shield; Gas shield; Tungsten electrode	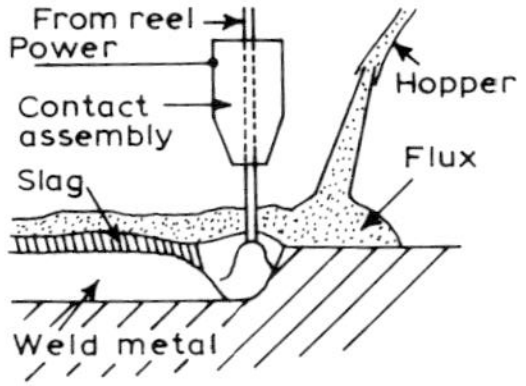 From reel; Power; Contact assembly; Slag; Hopper; Flux; Weld metal	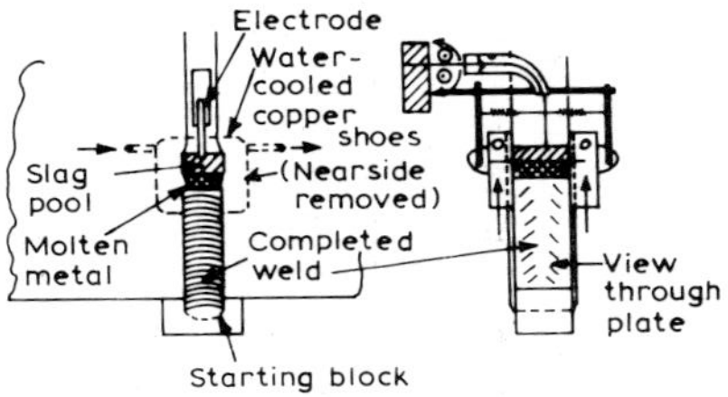 Electrode; Water-cooled copper shoes (Nearside removed); Slag pool; Molten metal; Completed weld; Starting block; View through plate
Typical applications	Low alloy steels Stainless steels Aluminium Copper Nickel Alloys (various)	Non-stainless steels Stainless (austenitic and ferritic) steels Aluminium Titanium Nickel Alloys etc	Production processes, e.g. high energy welding of large steel constructions. Suitable for steels, titanium, aluminium, etc.	Used for joining heavy castings and forgings. Suitable for steels, aluminium, titanium, etc
Metallurgical implications and problems	Porosity or cracking due to, e.g., absorption of nitrogen from air. Residual slag inclusions in weld	Metallurgically 'clean' process, i.e. low hydrogen and low inclusion content in weld metal. Porosity can arise because of CO_2 gas absorption (MIG) or due to inclusions (TIG)	High deposition rates possible. Porosity due to high cooling rates. Hot cracking problems when welding thick plates. Grain growth problems in HAZ due to high heat input	Leaves typical cast structure in weld (large columnar grains, dendrites, etc). Centre-line cracking (segregation). 'Undercut' problems. Porosity due to lack of fusion. Grain growth in the heat-affected zone (HAZ). Low hydrogen process. Low toughness (due to segregation)
Choice of slags	Typically: silica, alumina, rutile, magnesia, fluorspar, lime, manganese oxide, deoxidants (various)	No slags, but some deoxidants used	Similar to manual metal arc	Slags chosen to encourage high resistivity and blanket out air

plays an essentially different role from normal arc welding in that heat is generated by *resistance heating* of the slag. Thus, fluxes are chosen with respect to their effect on electrical conductivity and viscosity of the slag. In terms of heat input the electroslag process has a much larger heat capacity, and the heat source moves more slowly than in other fusion welding processes.

In order to appreciate the effect of the variables in fusion welding on the characteristics and properties of welded joints it is worth examining the principles of consumable development in more detail.

Principles of consumable development

Consumables in welding usually include a combination of wire and flux, although in, e.g., MIG and TIG (tungsten–inert gas) welding a flux is not employed, *Table 1.1*.

Consumables should possess the following characteristics (*see*, e.g., refs 4—6):

1. Provide good arc properties.
2. Give protection to metal drops during deposition.
3. Possess suitable physical properties, i.e. with respect to viscosity, melting temperature, surface tension and thermal expansion.
4. Be able to deoxidize the weld metal.
5. Be capable of dissolving solute gases.
6. Afford protection to the weld metal.
7. Assist in shaping the weld deposit.
8. Be able to transfer alloying elements to the weld metal.
9. Should be easy and cheap to manufacture.

There are three main types of consumable, these being:

1. Flux-covered, as used in MMA welding.
2. Flux-cored, used in various welding processes.
3. Fluxes for submerged arc welding.

The characteristics of these three types are briefly considered below.

MMA consumables As shown in *Figure 1.1* these consist of a solid wire of typical composition: 0.2 wt % C; 0.5 wt % Mn; 0.02 wt % Si. The flux coating comprises various types of powders (rutile, cellulose, basic and acid), depending on the properties required. For example, the use of high rutile (TiO_2) fluxes tends to improve the arc properties, while basic fluxes provide a cleaner weld deposit with lower inclusion content than a rutile flux. Some examples of these flux coatings, with comments on their properties, are given in *Table 1.2*.

Flux-cored electrodes These are based on a low carbon steel outer tube with a core of flux material. The advantages of these electrodes are:

1. Low hydrogen content because of good protection.

2. Assist alloying reactions, since flux material falls unmelted into the weld pool.

On the other hand, these consumables are not as easy or cheap to manufacture as conventional flux-covered wires.

TABLE 1.2. Some electrode coating types (after Gray, T.G.F., Spence, J. and North, T.H., *Rational Welding Design*, Butterworths, 1975)

Type	*Formulation*	*Comments*
Cellulosic	20—60% rutile, 10—20% cellulose, 15—30% quartz, 0—15% carbonates, 5—10% ferro-manganese	Cellulose promotes gas shielding in the arc region. Hydrogen increases heat at weld. Hydrogen content *high* (30—100 ppm). Deep penetration, fast cooling weld
Rutile	e.g. 40—60% rutile, 15—20% quartz, 0—15% carbonates, 10—14% ferromanganese, 0—5% organics	Slags mainly for slag shielding. Hydrogen content fairly high (15—30 ppm). High inclusion content in weld deposit
Acid	Iron ore–manganese ore, quartz, complex silicates, carbonates, ferromanganese	Fiarly high hydrogen content. High slag content in weld metal
Basic	20—50% calcium carbonate, 20—40% fluorspar, 0—5% quartz, 0—10% rutile, 5—10% ferro-alloys	Fairly low hydrogen levels ($\leq$10 ppm), hence commonly used in welding low alloy constructional steels. Electrodes should be kept dry. Low inclusion content in weld deposit

Submerged arc welding This is a high production process in which the flux powder is gravity fed, as shown in *Table 1.1*, which limits application to horizontal welding. The easier mixing and production of these flux powders allows their compositions to be determined directly from ternary and quaternary oxide diagrams or viscosity–temperature diagrams. The most common mixtures are based on $MnO–SiO_2$ and $CaO–SiO_2$, with additions of Al_2O_3, ZrO, TiO_2, etc. An example of a viscosity–temperature graph for typical submerged arc fluxes is shown in *Figure 1.2*. It is important in 'shaping' the weld that the slag solidifies *after* the weld metal, and *Figure 1.2* shows that slag viscosities do not increase significantly until temperatures fall below *ca.* 1300 °C.

Of the factors that govern weld metal composition, the basicity of the flux plays an important role. In general, the basicity B is expressed by equation

$$B = \frac{\Sigma \text{ (basic oxides)}}{\Sigma \text{ (acid oxides)}} \quad (1.1)$$

(1.1). In practice, the basicity is usually calculated in terms of the ratio between the basic oxides and SiO_2 (wt%), and the oxides are thus given in proportion to

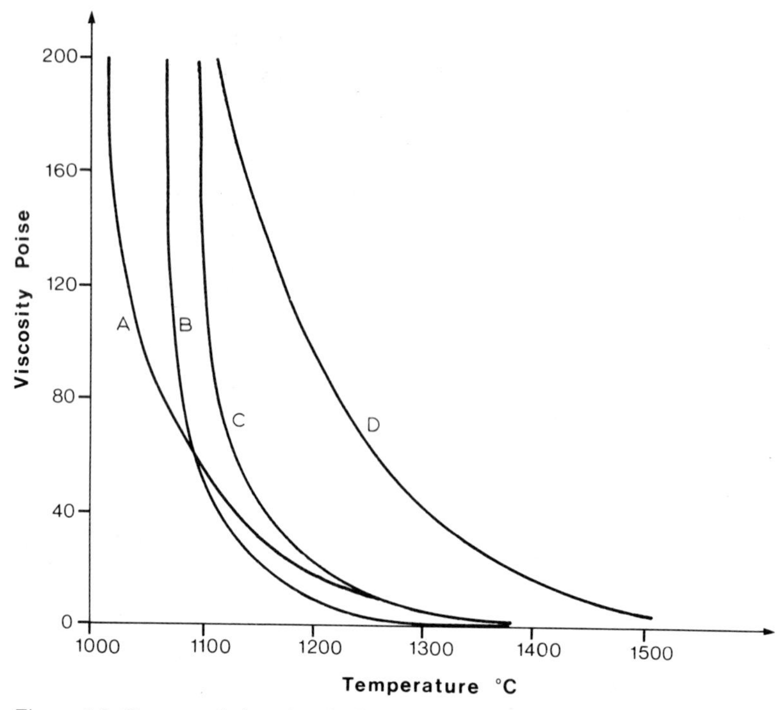

Figure 1.2 Changes of viscosity of submerged arc welding slags as a function of temperature. A, $CaO-MgO-Al_2O_3-SiO_2$; B, $MnO-SiO_2$; C, $CaO-SiO_2-TiO_2$; D, $CaO-SiO_2$. After Jackson, C., *Fluxes and Slags in Welding,* Ohio State University, Dec. 1973

their effect on the formation of monosilicates. Thus, a complete formulation may be given as in equation (1.2). Note that Al_2O_3 is considered here as basic in

$$B = \frac{\Sigma CaO}{\Sigma SiO_2}$$

$$= \frac{CaO + 1.4MgO + 0.60K_2O + 0.90Na_2O + 0.45ZrO_2\ (+\ 0.55Al_2O_3)}{SiO_2\ (+\ 0.59Al_2O_3)} \tag{1.2}$$

acid coatings and acid in basic coatings. Typically, B has values ranging from 0.5 to 3.0.

Examples of changes in slag basicity in MMA welds as a function of (*a*) the final weld composition and (*b*) the total inclusion content are given in *Figure 1.3.* As expected, the rutile fluxes tend to produce higher inclusion contents in the weld metal, with a resulting decrease in toughness. The importance of deoxidants based on Si is obvious in fusion welds. However, these oxidizing reactions are more difficult to control in welding than in steel making, with the result that weld deposits typically contain a fairly high density of silicates.

In some respects, fusion welding can be likened to a miniature electric steel making process. However, arc welding involves very short time scales and

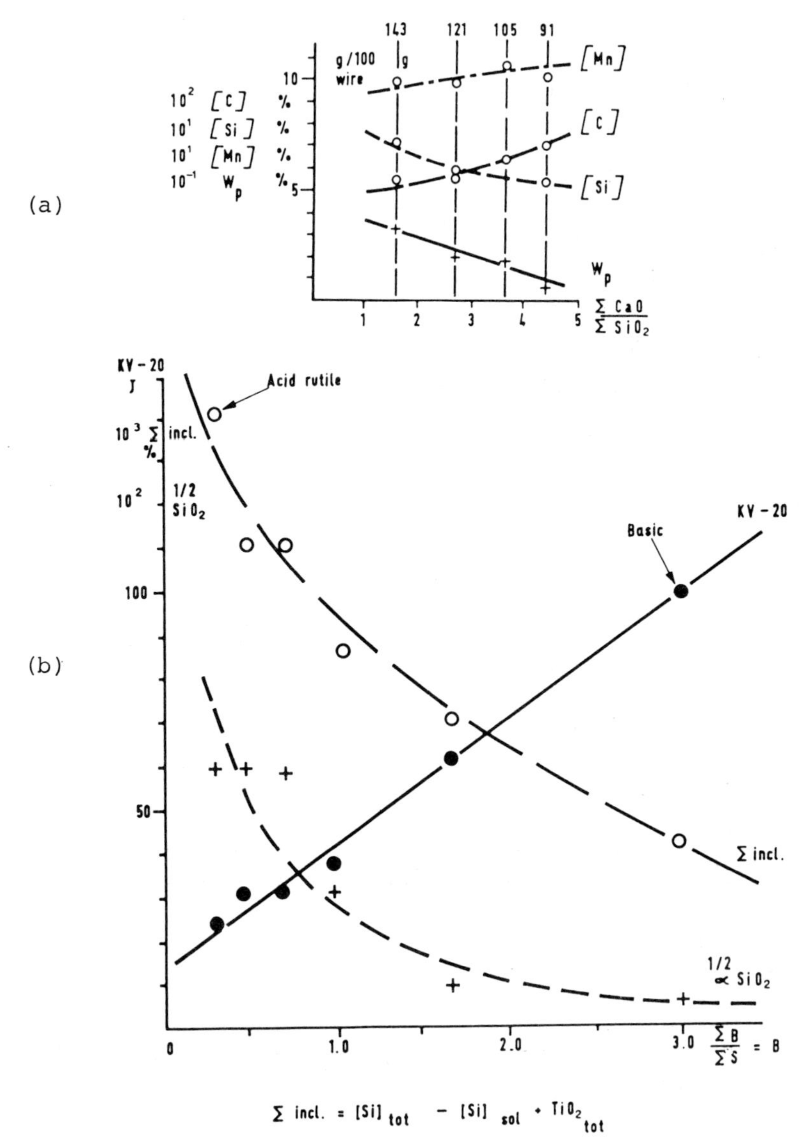

Figure 1.3 (a) Effect of basicity on the alloy content in the weld deposit. (b) The effect of basicity on the inclusion content and toughness of the weld deposit. Both figures after Almqvist *et al.*[5]

much higher temperatures than steel making, so that thermodynamic equilibrium is rarely achieved. An example of this departure from equilibrium in arc welding[6] concerns the decarburizing reaction (1.3) when using an

$$CO_2 + [C] \rightleftharpoons 2CO \qquad (1.3)$$

80% Ar – 20% CO_2 shielding gas. The square bracket indicates that carbon is in solution in the melt. It appears, in fact, that at carbon contents less than *ca.* 0.1% carburization takes place, which implies that dissociation of the CO_2 is controlled by the high temperature of the arc, not by the temperature of the melt. It can be shown on thermodynamic grounds that, at temperatures above *ca.* 3500 K, reaction (1.4) occurs. Indeed, the very steep temperature gradients

$$CO_2 \rightleftharpoons CO + \tfrac{1}{2}O_2 \quad (1.4)$$

associated with arc welding are thought to enhance turbulence of the weld pool, and this is likely to affect solidification behaviour (*see* Chapter 2).

The importance of keeping hydrogen content low in the weld metal has resulted in stringent control of the *hygroscopic* properties of fluxes. In this case, the flux should be resistant to moisture absorption during storage of the electrodes or flux powder, and the electrodes should be capable of being

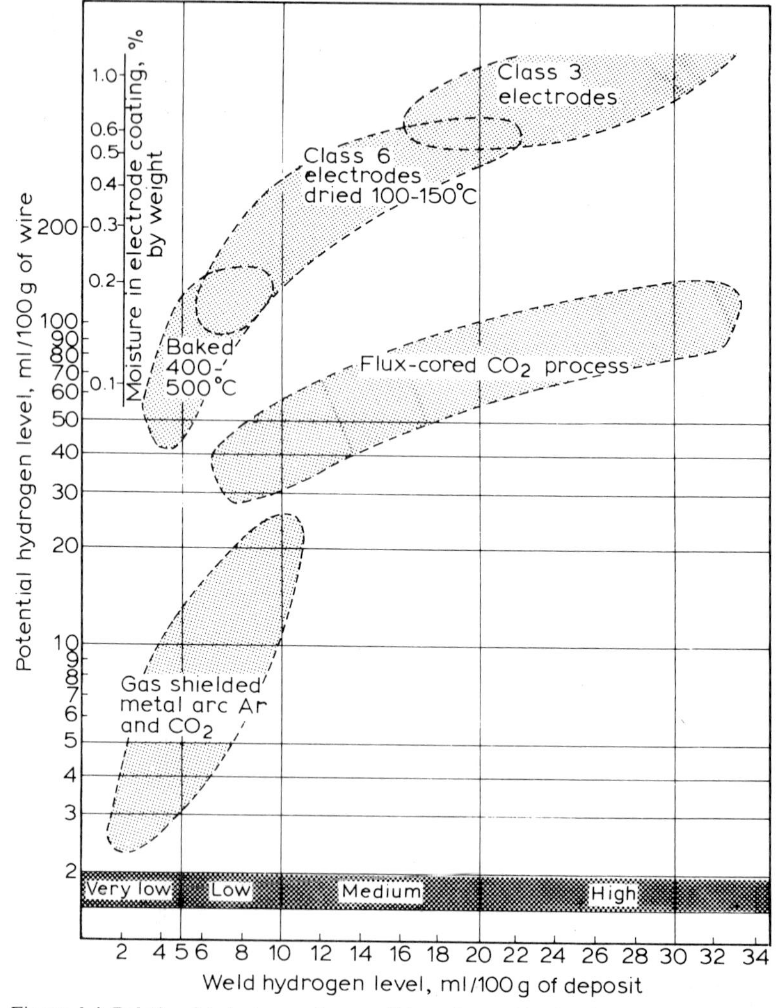

Figure 1.4 Relationship between the possible and actual hydrogen levels in welds as a function of different processes and drying procedures. After Coe, F.R., *Welding Steels without Hydrogen Cracking,* The Welding Institute, Abington, 1973

redried, if necessary, prior to use. As discussed in the excellent review of Coe[7], fused, submerged arc fluxes, being glasses, display low moisture levels as long as they are carefully dried after water quenching during manufacture. Agglomerated fluxes (as used in coated electrodes) contain, on the other hand, 4 — 7% water in the form of binder which is reduced to approximately 2% during drying. Of course, different formulations of flux absorb moisture at different rates, but in general rebaking at *ca.* 450°C can give a damp flux a low hydrogen content. Examples of the effect of baking temperature on the amount of residual hydrogen in the weld metal are shown in *Figure 1.4*.

Absorption of gases in the weld metal

It is seen from *Tables 1.1* and *1.2* that, in certain welding processes, gases can be absorbed into the weld pool and, if excessive, may give rise to porosity or cracking problems after solidification. For example, in spite of shielding, nitrogen and oxygen can be absorbed from the air. Hydrogen can be absorbed from moisture in the air, or through reactions that involve the fluxes, or even

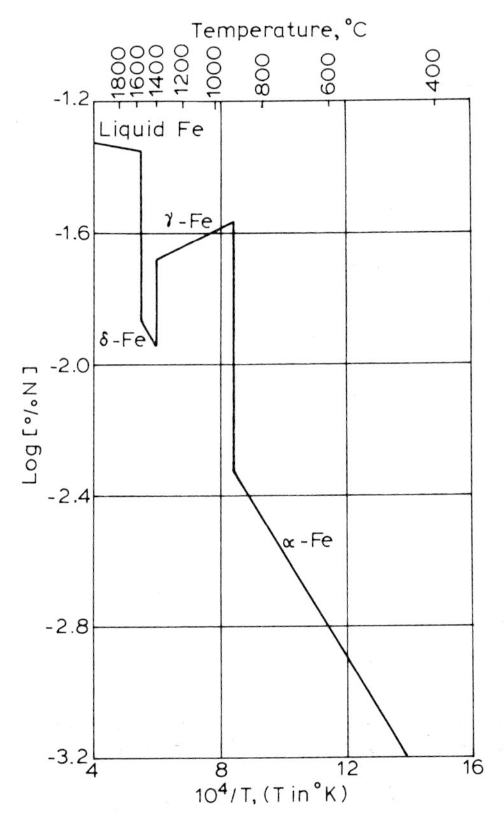

Figure 1.5 The solubility of nitrogen in iron as a function of temperature. After *The Making, Shaping and Treatment of Steel*, U.S. Steel, 1971

from surfaces which are not clean because of oil, dirt, grease or rust. As mentioned earlier, CO and CO_2 absorption may occur from the shielding gas itself. In most cases, nitrogen can be excluded fairly efficiently although problems can arise, particularly in the MMA technique. Small amounts of absorbed oxygen may actually be beneficial to the refining process, although particular care must be taken to avoid oxygen contamination when welding titanium and aluminium. However, any absorption of hydrogen is potentially very dangerous because of *cold cracking* problems, particularly when welding steels of high hardenability.

In view of the special significance of nitrogen and hydrogen absorption on the properties of welds, these are now considered further. For a more detailed survey *see*, e.g., Séférian[8].

Nitrogen absorption

Although air is the most important source of nitrogen, the constitution and thickness of the electrode coatings as well as the arc length may affect nitrogen content in the weld metal. It can either be absorbed as gas or in atomic form through reaction (1.5). Its equilibrium concentration in liquid iron at 1600 °C is

$$N_2 \rightarrow 2[N] \tag{1.5}$$

then about 0.0445 wt% at one atmosphere (*Figure 1.5*). This figure shows that the solubility of atomic nitrogen is considerably higher in face-centre cubic

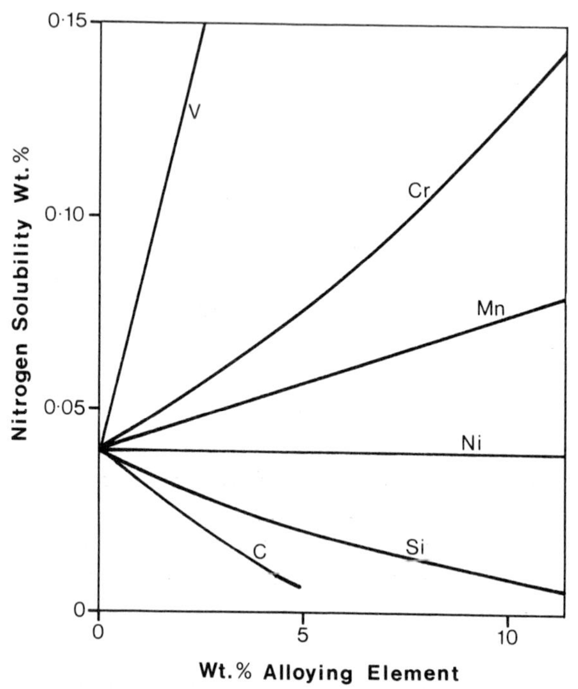

Figure 1.6 The effect of various alloying elements on the solubility of nitrogen in iron at 1600 °C. After Kumar, B., *Physical Metallurgy of Iron and Steel*, Asia Publications, 1968

(f.c.c.) γ–iron than in body-centre cubic (b.c.c.) δ– or α–iron. These solubility differences in the solid state can, of course, lead to the formation of various nitrides. The effect of other alloying elements present in liquid iron on the solubility of nitrogen is shown in *Figure 1.6.* It is seen that some elements effectively increase the solubility of nitrogen, although the presence of carbon, silicon and oxygen strongly reduce its solubility. These changes in solubility are, presumably, the result of misfitting strain effects and are assumed to apply to the solid state as well as to the liquid state.

In the essentially non-equilibrium conditions of welding, the actual amount of nitrogen found in solidified welds typically exceeds the values of *Figure 1.5.* This is shown in *Table 1.3,* where the nitrogen content of welds is

TABLE 1.3 Nitrogen content in various types of weld metal at ambient temperature (after Christensen, N., *Welding Compendium*, Norwegian Institute of Technology, Trondheim, 1975)

Weld	*Nitrogen content (wt%)*
Oxyacetylene welding, unalloyed steel	0.0016—0.002
Arc welding, unalloyed steel using:	
basic electrode	0.001
cellulose	0.0016—0.002
rutile	0.002 —0.0025
acid	0.0027—0.003
oxidizing	0.0035—0.004
uncoated	0.01 —0.022
Arc welding, 18–9 chrome–nickel steel, using:	
basic electrode	0.0055—0.0065
rutile	0.006 —0.0075
Submerged arc welding, melted powder (multi-run)	0.0014

given at ambient temperature as a function of different welding processes and conditions. Since the ambient solubility of atomic N in Fe is less than 0.001 wt% it is evident that the high values given in *Table 1.3* are due to the presence of nitrides, and that most of the excess of nitrogen derives from air.

Hydrogen absorption

Hydrogen absorption during welding can lead to various problems that cause, e.g., porosity in the solidified weld metal or cracking in the heat-affected zone. As with nitrogen, hydrogen readily decomposes in iron to its atomic state by reaction (1.6). The solubility of hydrogen in iron is shown in *Figure 1.7.* The

$$H_2 \rightarrow 2[H] \tag{1.6}$$

large decrease in solubility when the steel solidifies may cause porosity due to the reverse of reaction (1.6). The rapid diffusivity of atomic hydrogen also means that much of it diffuses out of the surface of the weld, although some inevitably penetrates deeper into the weld metal or base metal. The effect of alloying elements on the solubility of atomic hydrogen in iron melts at 1600 °C is shown in *Figure 1.8*. Note the marked decrease in solubility caused by the presence of carbon and boron, which has important implications in cold cracking problems (*see* Chapter 4).

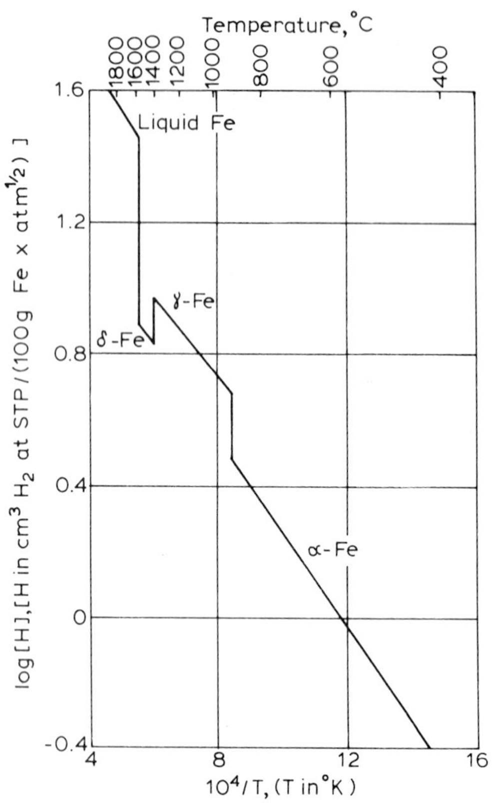

Figure 1.7 The solubility of hydrogen in iron. After *The Making, Shaping and Treatment of Steel,* U.S. Steel, 1971

Figure 1.9 shows how different types of welding process, cleanliness and dryness of electrodes may affect the absorption of hydrogen in the weld metal. *Figure 1.10* illustrates the distribution profiles of hydrogen in solidified weld metal as a function of distance from the weld crater for the cases of basic and rutile electrodes. These results emphasize the importance of using basic electrodes in conditions where cold cracking may be a problem, e.g. when welding thick, high strength steel sections. The significant reduction in hydrogen level behind the arc is mainly due to surface losses.

Composition of welds

Fusion welding is a process that involves a very intense heat source and this inevitably leads to melting back or *dilution* of the base metal. The weld metal

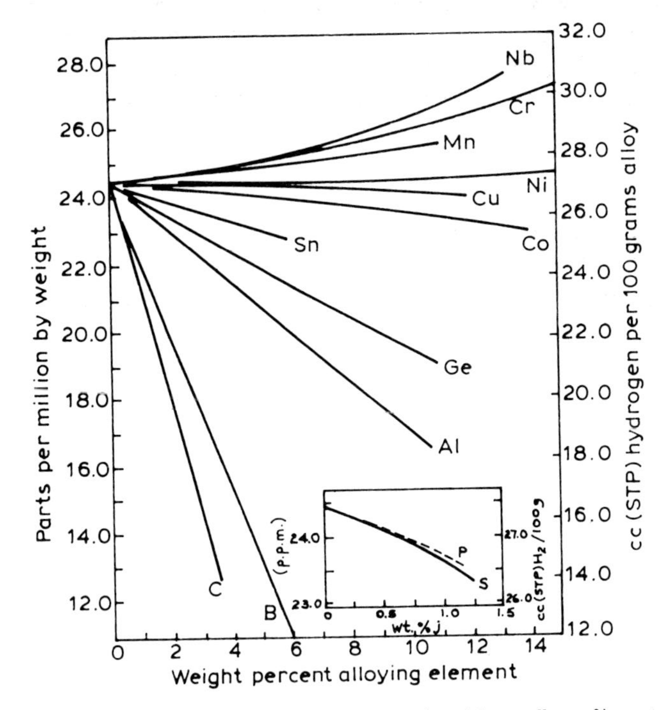

Figure 1.8 The solubility of hydrogen in various binary alloys of iron at 1592 °C. After *The Making, Shaping and Treatment of Steel*, U.S. Steel, 1971

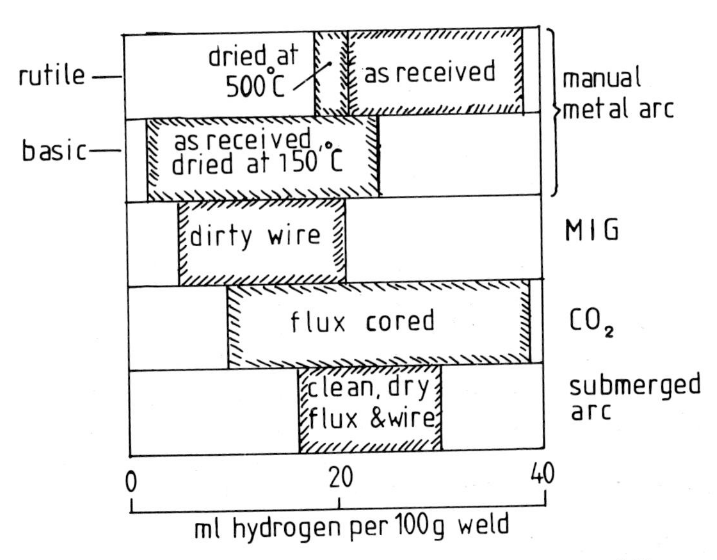

Figure 1.9 Weld metal hydrogen as a function of process variables; compare with *Figure 1.4*

thus consists of a mixture of several materials, obtained from the filler rod, the base metal and even the flux material.

To produce a good weld, considerable manipulation of the electrode is often necessary, as illustrated in *Figure 1.11.* The particular path of the electrode shown in this illustration avoids craters at the edges of plates, which can constitute high risk cracking areas. This is because the relatively shallow temperature gradient across the weld pool after the arc is removed increases the effect of segregation, and this, together with the residual stresses due to non-uniform cooling rates across the weld, may result in *solidification cracking.* These effects are discussed in more detail in Chapter 4.

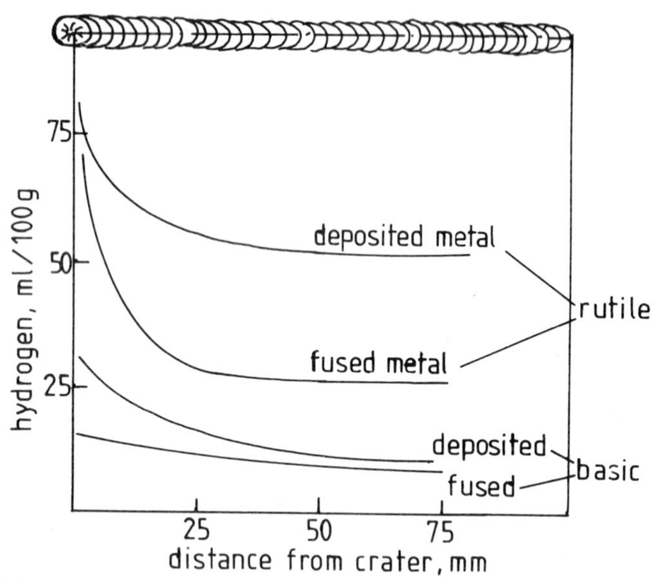

Figure 1.10 Measured hydrogen distributions as a function of type of electrode and distance from the weld crater. After Christensen, N., *Welding Compendium,* Norwegian Institute of Technology, Trondheim

In the case of thick plate, multi-run welds, the arc path has to be carefully controlled to bring about a neat, complete weld as shown, for example, in a single vee-butt weld (*Figure 1.12(a)*) and a vertical butt weld (*Figure 1.12 (b)*). Multi-run welds such as these are clearly more time consuming and costly than single deposit welds, but may be necessary in order to reduce *residual stresses* in critical cases such as welding nuclear pressure vessels or pipe lines. Cross-sections of MMA multi-run welds are shown in *Figure 1.13* to illustrate (*a*) a single vee-butt weld and (*b*) a double vee-butt weld. These micrographs show how the original cast structure of the weld metal is locally reheated and modified in appearance and this effect is discussed in more detail in Chapter 3. Note also that these micrographs show how even carefully deposited welds leave surface undulations, which may, in some cases, affect fatigue strength. In certain applications the first weld strings, at the centre of the double vee, may

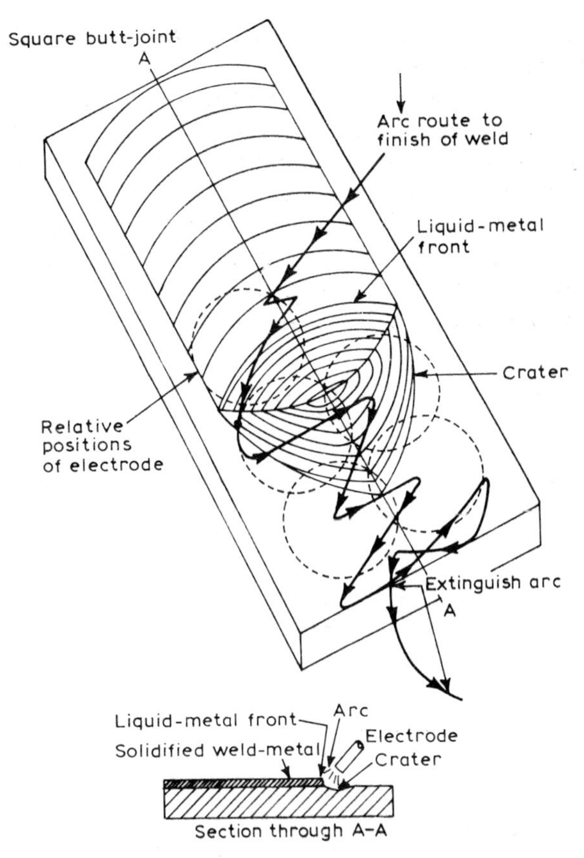

Figure 1.11 Illustrating how end cracking can be avoided by appropriate manipulation of the electrodes. After Woods, P.F., *Fundamentals of Welding Skills,* Macmillan, London, 1976

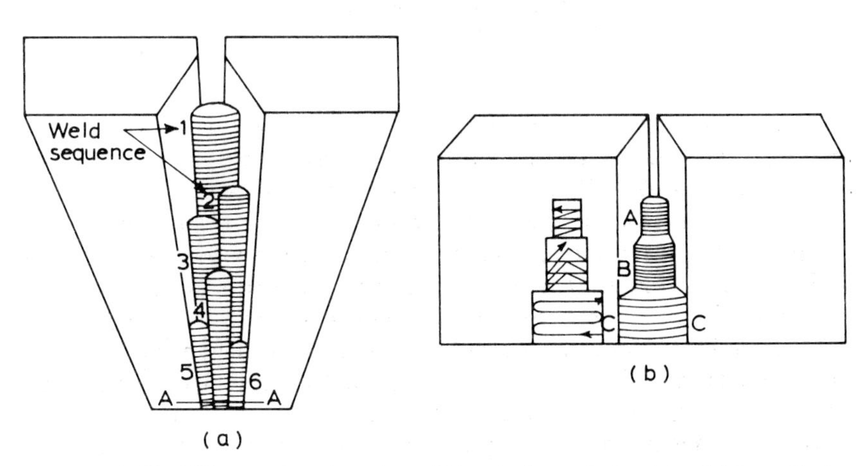

Figure 1.12 Ideal filler metal configurations of (a) a single vee-butt weld and (b) a vertical butt weld. After Woods, P.F., *Fundamentals of Welding Skills,* Macmillan, London, 1976

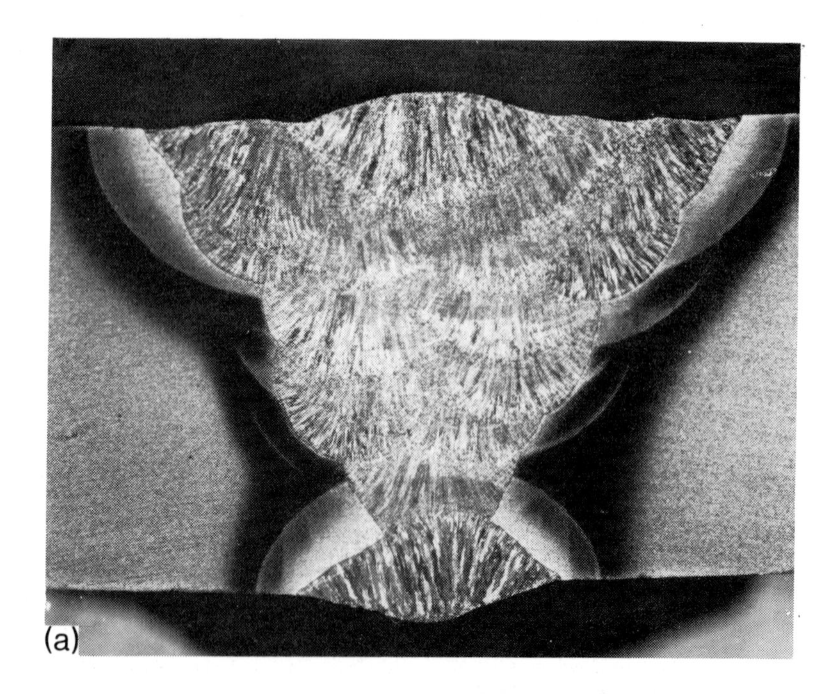

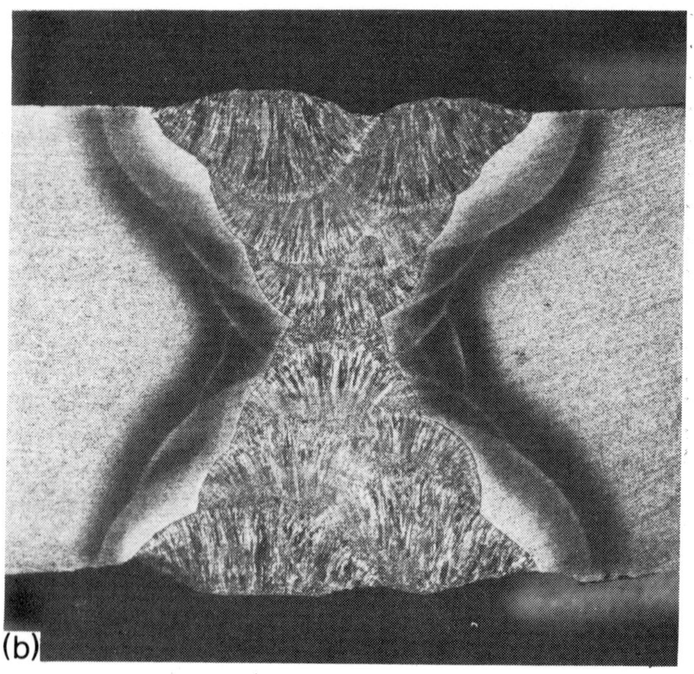

Figure 1.13 Manual metal arc welds of (a) a single vee-butt weld and (b) a double vee-butt weld. The plate thickness in both cases is 18 cm. (By courtesy of ESAB, Gothenburg)

be of a softer material than the other weld strings, in order to reduce residual stresses. This technique, known as 'buttering', is used in certain critical applications, e.g. to avoid *lamellar tearing* (*see* Chapter 4).

Summary of process variables

Fusion welding is a rather complex process of many variables, any of which may have an important effect on the final microstructure and properties of the welded joint. It seems to be difficult to prevent gases penetrating the melt, particularly oxygen, nitrogen and hydrogen, and of these hydrogen is potentially the most problematical. It is thus necessary to consider the role of hydrogen in detail (Chapter 4). The flux materials in coated electrodes or in submerged arc welding have important effects on the final weld properties because of alloying and residual inclusions. The composition of the weld metal is also bound to be affected by the composition of the base metal because of dilution, and in some cases this effect is also detrimental to weld properties (*see* Chapter 2).

Certain modifications to the cast structure of weld metal can be induced using multi-run welding, and this may be important in thick welded joints subjected to high residual stresses. Indeed, the severity of the intense heat source used in welding implies that very steep temperature gradients are generated around the melt zone. The weld thermal cycle is considered in more detail in the next section.

The weld thermal cycle

Arc welding is a process in which a very intense, moving heat source is applied to the workpiece. It would be extremely useful to predict the form of the temperature gradients about this heat source in order to understand such phenomena as the width and penetration depth of a weld as a function of a given heat input, microstructural changes in the heat-affected base metal, and residual stresses. Knowledge of all these conditions is also necessary if predictions are to be possible concerning cracking problems.

Experimentally, it is difficult to make measurements of the temperature distribution within the weld pool, although a number of measurements of the weld thermal cycle in the heat-affected base metal have been made using embedded thermocouples; these results are considered later. Theoretical analyses of the weld thermal cycle have also been attempted, although these usually require a number of simplifying assumptions as regards, e.g., the material properties of the melt and even the high temperature properties of the base metal. Nevertheless, as is shown below, rather satisfactory agreement has been obtained between measured and theoretical temperature profiles in the heat-affected zone.

Heat-flow equations

Solutions to the heat-flow equations of a moving point heat-source were given by Rosenthal[9] in 1935. He assumed that the energy of the heat source moves with a constant speed v along the x axis of a fixed rectangular co-ordinate system, as shown in *Figure 1.14* and equation (1.7). It is usual to define the heat

$$\text{Heat flux} = q/v \text{ MJ m}^{-1} \text{ of weld} \qquad (1.7)$$

flux, q, in terms of the welding current and voltage (I and V) and the efficiency of the arc, η, equation (1.8). The arc efficiency depends on the welding process

$$q = \eta VI \qquad (1.8)$$

and some typical values are given in *Table 1.4*. The 1—10% loss in efficiency of the submerged arc process is related to that part of the flux which does not fuse. The low efficiency of the TIG process is associated with heat losses through the electrode holder.

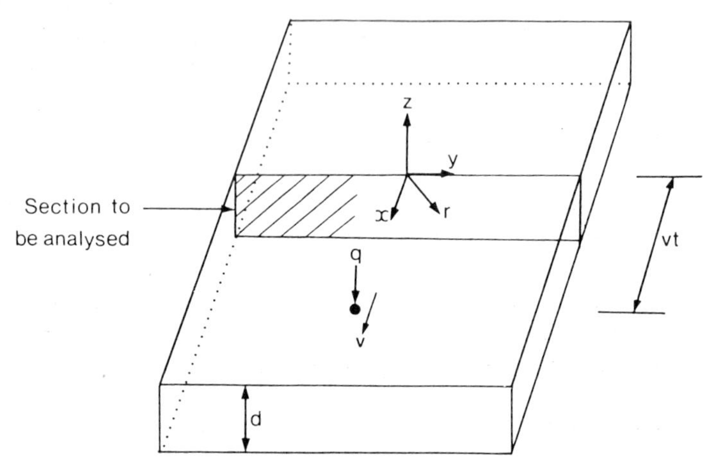

Figure 1.14 Welding configuration in terms of a point heat source, q, and a constant velocity, v

Equation (1.9) is the differential equation of heat flow expressed in the co-ordinates given in *Figure 1.14*, where T is the temperature (K), t is the time

$$\frac{\partial^2 T}{\partial x^2} + \frac{\partial^2 T}{\partial y^2} + \frac{\partial^2 T}{\partial z^2} = 2\lambda \frac{\partial T}{\partial t} \qquad (1.9)$$

(s) and λ is the thermal conductivity (J m^{-1} s^{-1} K^{-1}). Note that this approach assumes that the thermal conductivity and the specific heat × density product

TABLE 1.4 Arc efficiencies

Process	η
Manual metal arc	0.7 —0.85
Tungsten–inert gas (argon)	0.22—0.48
Metal–inert gas (argon)	0.66—0.75
Submerged arc	0.90—0.99

(ϱc) of the base metal are constants. This is not correct, of course, and possible implications of this assumption are discussed later. Equation (1.9) essentially refers to a fixed co-ordinate system, and may be modified to a moving co-ordinate system by replacing x with ξ, where ξ is the distance of the point heat source from some fixed position along the x axis, with dependence on the velocity of the moving source, to give equation (1.10). Equation (1.9), when

$$\xi = x - vt \tag{1.10}$$

differentiated with respect to ξ, gives equation (1.11) which can be simplified if

$$\frac{\partial^2 T}{\partial \xi^2} + \frac{\partial^2 T}{\partial y^2} + \frac{\partial^2 T}{\partial z^2} = -2\lambda v \frac{\partial T}{\partial \xi} + 2\lambda \frac{\partial T}{\partial t} \tag{1.11}$$

a so-called *quasi-stationary* temperature distribution exists. This means that the temperature distribution around a heat source of uniform velocity will settle down to a constant form, i.e. $\partial T/\partial T = 0$ provided $q/v =$ a constant. Since this ought to be achieved in most welding situations, equation (1.11) can be simplified to give equation (1.12). Consider the dimensions of the plate being

$$\frac{\partial^2 T}{\partial \xi^2} + \frac{\partial^2 T}{\partial y^2} + \frac{\partial^2 T}{\partial z^2} = -2\lambda v \frac{\partial T}{\partial \xi} \tag{1.12}$$

welded. In the case, for example, of a bead-on-plate weld, the flow of heat from the source depends on the thickness of the plate. This is illustrated in *Figure 1.15* for thick and thin plates. In effect, the flow of heat is two-dimensional for the thin plate and three-dimensional for the thick plate. Heat losses through the surface are thus assumed to be negligible, although this assumption is not always valid, particularly when welding thin plates.

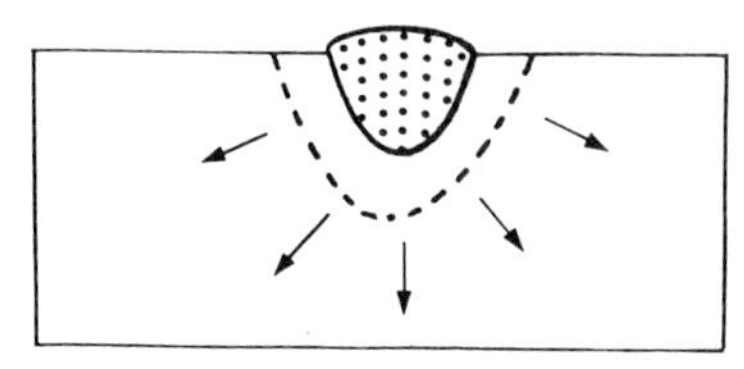

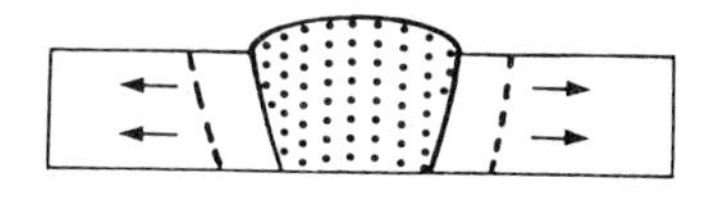

Figure 1.15 Three- and two-dimensional heat flow in welding

Equation (1.12) can now be solved for both thick and thin plates. Equation (1.13) is the solution for three-dimensional heat flow (thick plate) and equation

$$T - T_0 = \frac{q/v}{2\pi\lambda} \exp\left(\frac{-v\xi}{2a}\right) \frac{\exp(-vr/2a)}{r} \tag{1.13}$$

(1.14) that for two-dimensional heat flow (thin plate), where a, the thermal

$$T - T_0 = \frac{q/v}{2\pi\lambda d} \exp(v\xi/2a) K_0(vr/2a) \tag{1.14}$$

diffusivity, is given by $a = \lambda/\varrho c$ m^2 s^{-1}, K_0 is a Bessel function of the first kind, zero order, $r = (\xi^2 + y^2 + z^2)^{1/2}$ and ϱc is the specific heat per unit volume in J m^{-3} K^{-1}.

Application of these equations to the thermal cycle experienced by the heat-affected zone (assuming instantaneous application of heat) at a certain fixed position from the heat source as defined by a radial distance r (given above) gives, for a thick plate, equation (1.15). Note that since t occurs in this equation

$$T - T_0 = \frac{q/v}{2\pi\lambda t} \exp\left(-\frac{r^2}{4at}\right) \tag{1.15}$$

(compare with equation (1.13)), the volume element in the HAZ, as defined by r, experiences the thermal cycle as a function of temperature and time. This is an important result and provides the basis for much of the following considerations. T_0 in equation (1.15) refers to the initial temperature prior to welding; T_0 may also refer to the *pre-heat temperature,* in K. It should be noted that equation (1.15) is essentially a first-order solution to equation (1.12) and is strictly to be applied only at temperatures *outside* the fusion zone[9].

Similarly, for the thin plate condition, the temperature – time distribution is given by equation (1.16), where d is the thickness of the plate.

$$T - T_0 = \frac{q/v}{d(4\pi\lambda\varrho ct)^{1/2}} \exp\left(-\frac{r^2}{4at}\right) \tag{1.16}$$

The form of the temperature distribution fields given by equations (1.13) and (1.14) is illustrated in *Figure 1.16.* Note that the intense heat source and very effective heat sink of the plates being welded result in an extremely steep temperature profile. The movement of the arc results in a bow-wave effect in which the isotherms effectively pile up at the leading edge. Comparisons between the welding of thick and thin plates are illustrated in *Figures 1.17* (*a*) and (*b*).

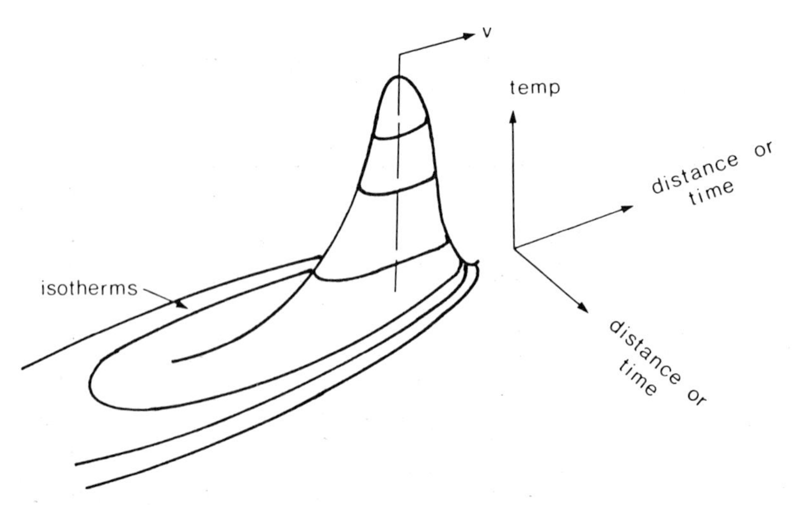

Figure 1.16 Three-dimensional temperature distribution in arc welding

The curves $n - n$ in these figures represent the locus of the boundary between rising and falling temperature in the solid. Thus, all points to the left of $n - n$ are on the cooling cycle, and points to the right are on the heating cycle. It can be shown that the effect of increasing welding speed is to concentrate the heat closer to the source and this has important implications on the solidification structure and properties of the weld (*see* Chapter 2).

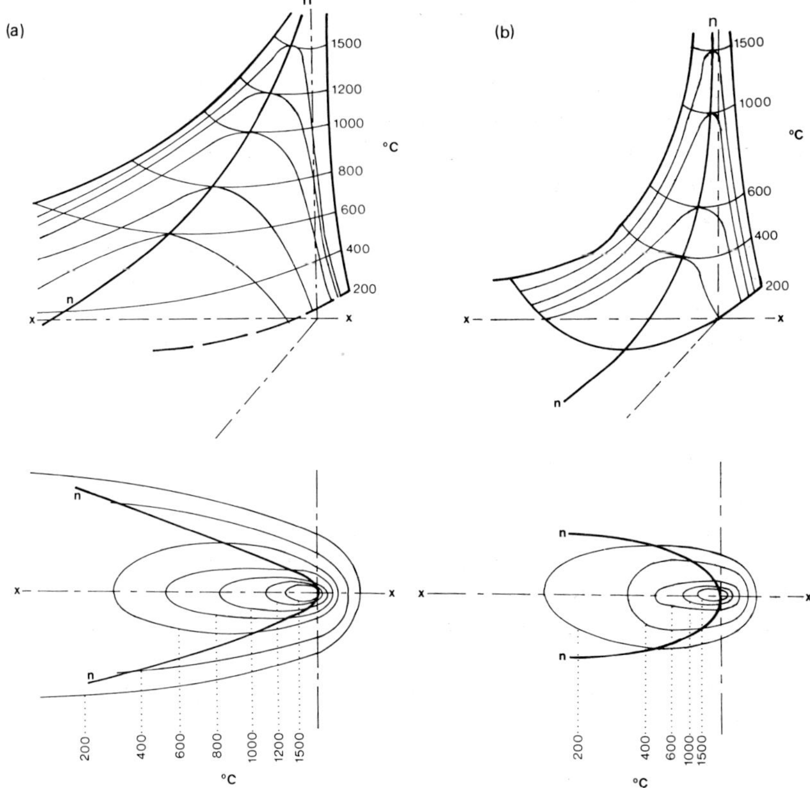

Figure 1.17 Calculated temperature distributions for (a) thin plates and (b) thick plates. The material being welded is mild steel. After Rosenthal[9]

TABLE 1.5 Material properties*

Material	*Volume thermal capacity* ϱc $(J\,m^{-3}\,K^{-1})$	*T–Diffusivity* $a\ (m^2\,s^{-1})$	*T–Conductivity* $\lambda\ (J\,m^{-1}\,s^{-1}\,K^{-1})$	*Melting point* (K)
Aluminium	2.7×10^6	$8.5\text{—}10\times10^{-5}$	229	933
Carbon steel	4.5×10^6	9.1×10^{-6}	41.0	1800
9% Ni steel	3.2×10^6	1.1×10^{-5}	35.2	1673
Austenitic steel	4.7×10^6	5.3×10^{-6}	24.9	1773
Inconel 600	3.9×10^6	4.7×10^{-6}	18.3	1673
Ti alloy	3.0×10^6	9.0×10^{-6}	27.0	1923
Copper	4.0×10^6	9.6×10^{-5}	384.0	1336
Monel 400	4.4×10^6	8.0×10^{-6}	35.2	1573

* Data from Gray, T.G.F., Spence, J. and North, T.H., *Rational Welding Design*, Newnes–Butterworths, 1975

An increase in current density extends the range of the isotherms, but does not affect their shape. An increase in conductivity and diffusivity of the metal affects both the shape and range of the isotherms. This implies that the shape and range of the isotherms are highly dependent upon the material being welded. Examples of the thermal properties of various alloys are shown in *Table 1.5*. For a given heat input, thick plates are more efficient heat sinks than thin plates and the isotherms are modified accordingly (*Figure 1.17* (*a*) and (*b*)). It should also be noted that if preheat is applied, the Rosenthal equations predict an effective numerical increase in the value of a given isotherm, but do not change its shape or range. This result is an important one and is discussed again in Chapter 3.

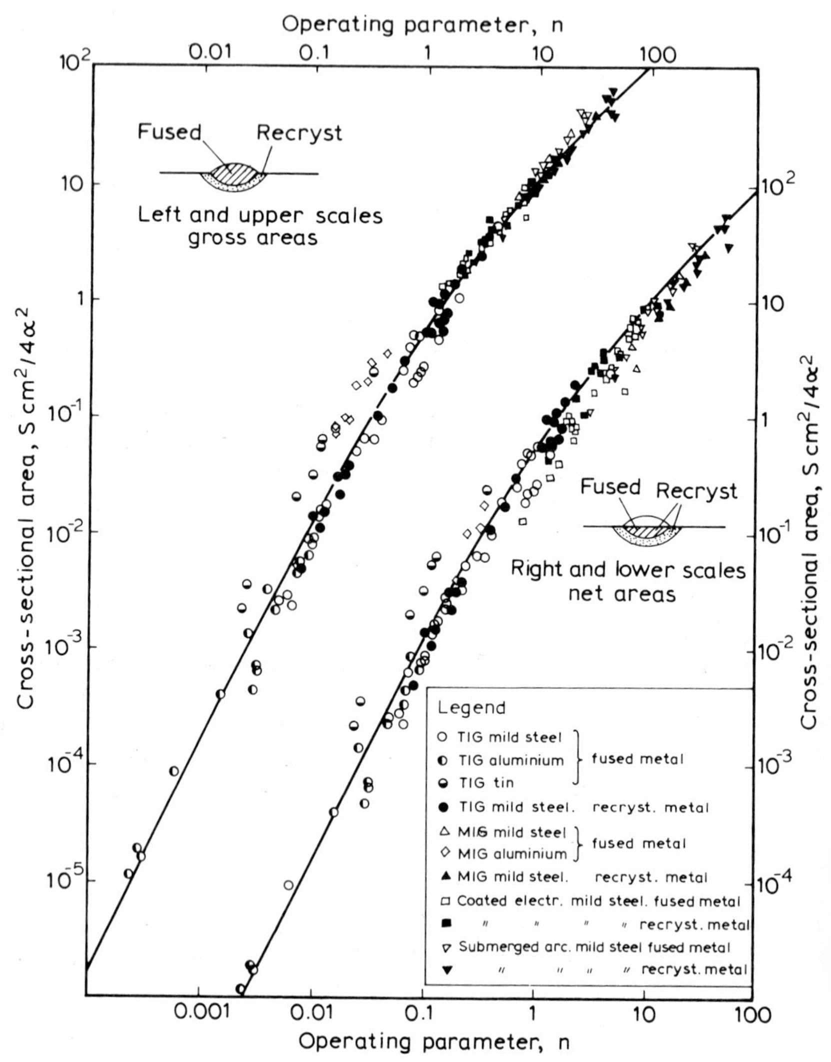

Figure 1.18 Comparison between predicted and measured weld cross-sections; n is a dimensionless parameter that allows various types of weld (and material) to be compared on a single diagram. After Shinoda and Doherty[10]

Attempts have been made to use the Rosenthal equations to help predict the size and shape of the weld deposit as a function of different types of welding (for a useful review of this approach as well as other empirical approaches to this problem, *see* Shinoda and Doherty[10]). Fairly satisfactory predictions of cross-sections of fused and recrystallized metal have been accomplished. An example of this work is shown in *Figure 1.18* in which comparisons between predicted (drawn curves) and experimental results for a number of welding process variables are shown.

In summary, considering the assumptions involved, e.g. point heat source, constant thermal properties, etc, there is good evidence of the general validity of the Rosenthal equations in describing the temperature distribution about a weld. As far as weldability is concerned, this is an important result since predictions of how the weld thermal cycle affects the microstructure of the base metal for a given welding process should be possible.

The thermal cycle of the base metal

It is an interesting fact that in most practical welding situations (a possible exception is electroslag welding, *see later*), the velocity of the arc along the workpiece is much higher than the velocity of thermal diffusivity. In other words, heat flow in the direction of travel is small compared to that perpendicular to the travel direction. This implies that a given slice of the material of the base plate 'feels' the weld cycle as a short, intense heat pulse of linear power q/v. The time taken to dissipate this heat pulse within the slice in thick plates is proportional to the thermal conductivity, λ, (equation (1.15)), while for thin plates, it is proportional to both λ and the specific heat per unit volume of the base metal, ϱc. On this basis, the thermal cycle for a given point in the base metal can be represented by a simple temperature–time diagram. From experimental measurements of the weld thermal cycle it has been found that, for a given welding process, weld geometry and material, the cooling time through the range 800—500 °C is constant, at least within the heat-affected base metal (at a peak temperature, T_p>900 °C), i.e. $q/v \propto \Delta t_{8-5}$ = const. The temperature of 800 °C is important in that in most steels it approximately represents the A_3 transformation temperature. Indeed, the use of Δt_{8-5} as descriptive of a given weld has now been widely adopted in welding circles. On this basis, equations (1.15) and (1.16) may be simplified considerably. In the welding of thick plates, following from equation (1.15), equation (1.17) can

$$T_p - T_0 = \left(\frac{2}{\pi e}\right) \frac{q/v}{\varrho c r^2} \tag{1.17}$$

now be written where T_p refers to the peak temperature of the thermal cycle and e is the base of natural logarithms (= 2.718). The cooling time, Δt, is given

$$\Delta t_{8-5} = \frac{q/v}{2\pi\lambda\Theta_1} \tag{1.18}$$

by equation (1.18), where Θ_1 is defined by equation (1.19). Equation (1.20) can be derived from these equations and it shows that the quantities T_p, T_o and

$$\frac{1}{\Theta_1} = \left(\frac{1}{773 - T_0} - \frac{1}{1073 - T_0}\right) \quad (1.19)$$

$$T - T_0 = \Theta_1 \frac{\Delta t}{t} \exp - \left(\frac{\Delta t}{et}\left(\frac{\Theta_1}{T_p - T_0}\right)\right) \quad (1.20)$$

Δt completely characterize the weld cycle. For the thin plate case equations (1.21) and (1.22) hold, where Θ_2 is given by equation (1.23). Hence equation (1.24) is obtained.

$$T_p - T_0 = \left(\frac{2}{\pi e}\right)^{1/2} \frac{q/v}{d\varrho c\, 2r} \quad (1.21)$$

$$\Delta t = \frac{(q/v)^2}{4\pi\lambda\varrho c\, \Theta_2{}^2\, d^2} \quad (1.22)$$

$$\frac{1}{\Theta_2{}^2} = \frac{1}{(773 - T_0)^2} - \frac{1}{(1073 - T_0)^2} \quad (1.23)$$

$$T - T_0 = \Theta_2 \left(\frac{\Delta t}{t}\right)^{1/2} \exp - \left(\frac{\Theta^2{}_2\, \Delta t}{2et\,(T_p - T_0)^2}\right) \quad (1.24)$$

Using these equations, a temperature–time profile for a thin plate model of the submerged arc welding of a Nb-microalloyed steel with $T_p = 1150$ °C and $\Delta t_{8-5} = 40$ s is compared in *Figure 1.19,* with a measured thermal cycle. Comparison between the two curves shows that the results agree fairly well,

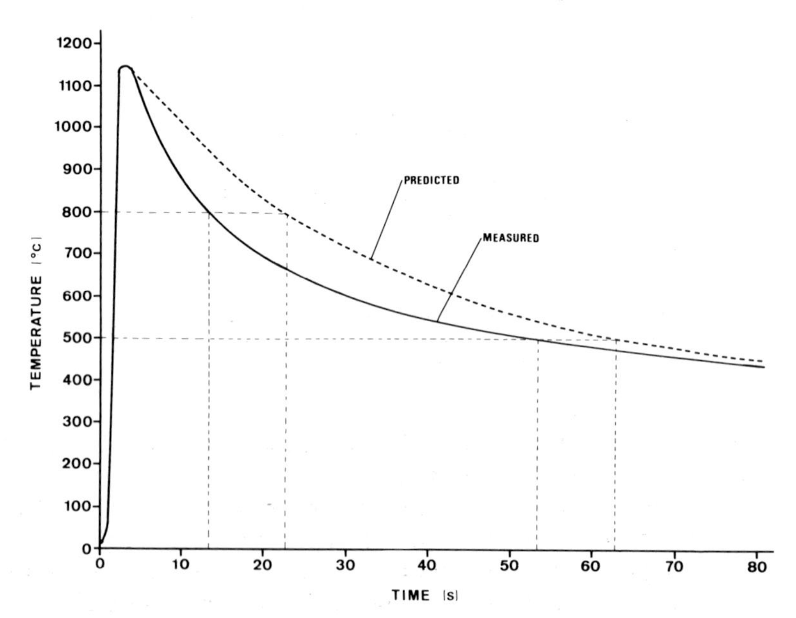

Figure 1.19 Comparison between measured and predicted thermal cycles in a Nb-microalloyed steel. A thin plate is assumed. Courtesy of J. Ion and J. Strid, University of Luleå

although the measured initial decrease in temperature is more rapid than predicted.

Clearly, the choice of equation used is governed not only by plate thickness, but by the weld process used and even the type of material. By equating the thin and thick plate approaches, a critical thickness, d', which defines the cross-over or boundary condition between the two can be derived, equation (1.25). The effect of the different variables q, T_o, etc, on Δt_{8-5} as a

$$d' = \left\{ \frac{q}{2\varrho c v} \times \left(\frac{1}{773 - T_0} + \frac{1}{1073 - T_0} \right) \right\}^{1/2} \tag{1.25}$$

function of d is shown in *Figure 1.20;* the dotted line represents d'. As expected, above a certain plate thickness (depending on input energy and preheat temperature used) the heat flow due to welding is essentially three-dimensional, and thus independent of thickness.

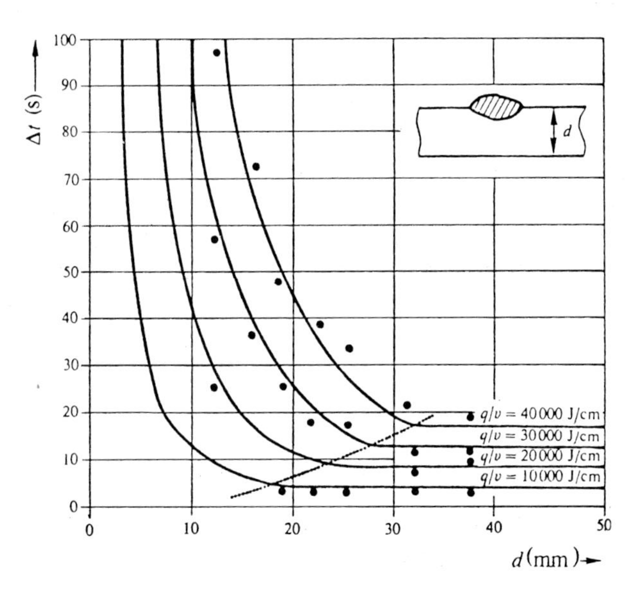

Figure 1.20 Calculated and measured values of Δt *vs.* plate thickness for various heat inputs. After van Adrichem, J. and Kas, J., Smitweld (weld simulator brochure), Holland

It is important to note that Δt_{8-5} effectively represents the cooling time for a given heat input, material type, weld geometry, etc. It is sometimes more convenient to express the weld cycle in terms of the heat input (q/v) and plate thickness d. Since Δt_{8-5} is directly proportional to q/v, it is a simple matter to derive Δt by graphical methods, and these graphs are usually referred to as *nomograms*. An example of such a nomogram is shown in *Figure 1.21*. In this example, Inagaki and Sekiguchi[11] have derived a modified form of equations (1.18) and (1.22); namely equation (1.26) which describes CO_2/O_2 arc welding.

$$\Delta t_{8-5} = \frac{(q/v)^{1.7}}{1.11 \times 10^6 \times \beta\left\{1 + \frac{2}{\pi}\tan^{-1}\left(\frac{d-13}{3.5}\right)\right\}} \tag{1.26}$$

In this equation, β is a constant, equal to 1 for a bead-on-plate weld or for when the final layer of a butt weld is laid, and equal to 1.7 when a T-fillet joint is welded. T_0 in this case is in °C. Thus, *Figure 1.21* can be used to compare the bead-on-plate and T-fillet welds. On this basis, a straight line between heat input and plate thickness in the nomogram intersects the appropriate Δ*t*.

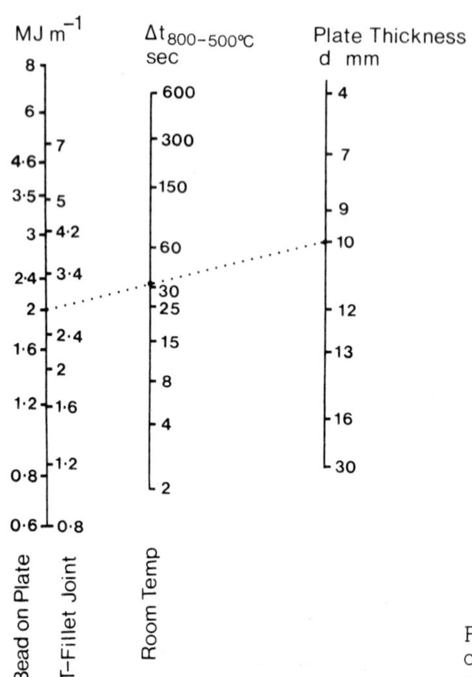

Figure 1.21 Nomogram for calculating the cooling time in the case of CO_2/O_2 arc welding. After Inagaki and Sekiguchi[11]

Refinements to the heat-flow equations

To a first approximation at least, the Rosenthal equations give a reasonable description of the thermal cycle as experienced by the heat-affected base metal. However, it is important to be aware of the assumptions made by Rosenthal in his approach, and how these assumptions may affect certain applications of the theory. The most important assumptions are:

1. A moving point source (or line) represents the weld arc.
2. Material properties are unaffected by changes in temperature or even of state (solid or liquid).
3. Heat losses at the plate surface are negligible.
4. The latent heat change due to phase transformations is neglected.

Examples of changes in thermal diffusivity as a function of temperature are given in *Figure 1.22,* which shows that drastic changes in iron-based alloys are likely to occur as a result of phase transformations. However, it greatly complicates the heat-flow equations to take these additional variables into account, and numerical techniques such as *finite-element analysis* (FEM) are needed to solve the equations. The advantage of using FEM, however, is that it is a convenient method for comparing heat-flow and thermal stress solutions,

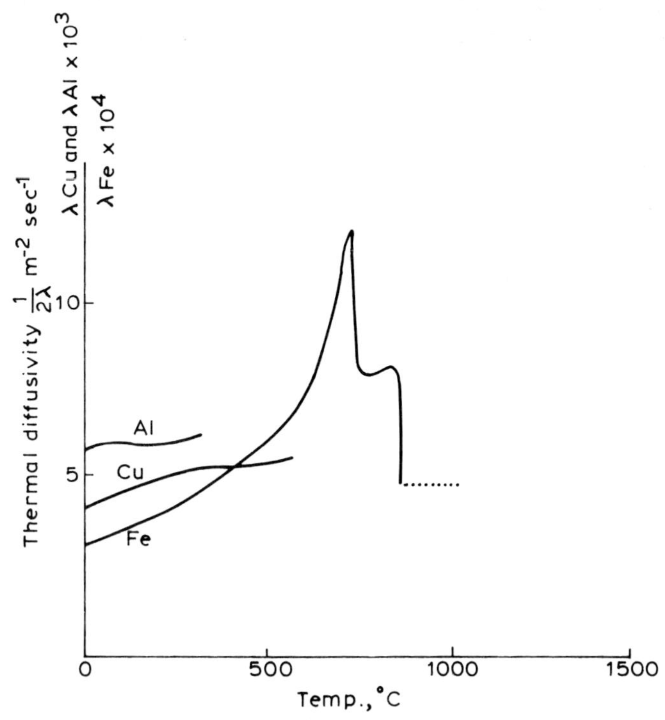

Figure 1.22 Variations of thermal diffusivity with temperature. After Rosenthal[9]

and this is important when calculating *residual stresses* in welded joints. On this basis, the complete heat-flow equation for a cubic lattice is expressed[12] by equation (1.27), where λ now refers to the thermal conductivities in three

$$\lambda(T)\left\{\frac{\partial^2 T}{\partial x^2}+\frac{\partial^2 T}{\partial y^2}+\frac{\partial^2 T}{\partial z^2}\right\}+\frac{\partial \lambda}{\partial T}\left\{\left(\frac{\partial T}{\partial x}\right)^2+\left(\frac{\partial T}{\partial y}\right)^2+\left(\frac{\partial T}{\partial z}\right)^2\right\}+Q-\varrho c(T)\frac{\partial T}{\partial t}=0 \tag{1.27}$$

mutually perpendicular directions, as defined in *Figure 1.14,* and Q is the *external* heat input per unit volume, whether it be a heat sink or heat source. For a discussion of the boundary conditions and application of equation (1.27), *see,* e.g., ref. 12. Comparison between equations (1.27) and (1.12) shows that, unlike equation (1.12), equation (1.27) avoids assumptions 2, 3 and 4 above.

In practice, FEM solutions of equation (1.27) have been attempted by assuming that the temperature field in the molten metal is governed by the same equation as in solid metal[13]. Energy transfer from the electrode to the workpiece is simulated by surface heat sources. The stirring effect in the molten metal is simulated by using a large value of thermal conductivity . By using these additional refinements it is claimed that improvements in the comparison between theoretical and experimental weld thermal cycles are achieved. In particular, the effect of the $\delta \rightarrow \gamma$ and $\gamma \rightarrow \alpha$ transformations with their associated latent heat changes, have been shown to produce slight 'kinks' on the thermal cycle curves, and these kinks are simulated by using refined

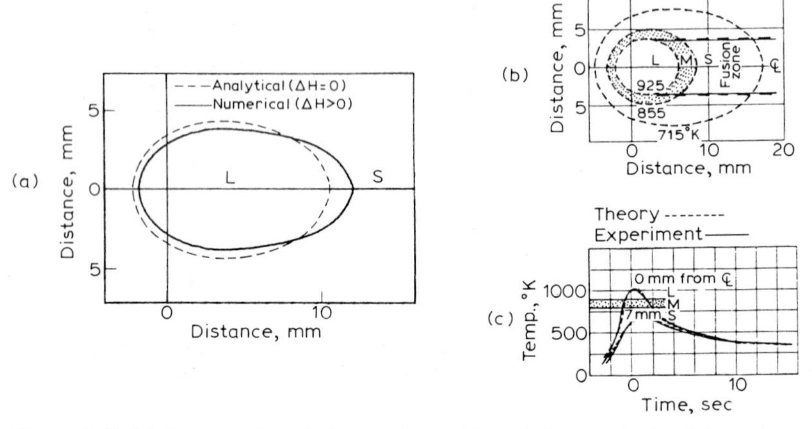

Figure 1.23 (a) A comparison between the weld pool shape calculated from the analytical solution to the heat-flow equation of Rosenthal (dotted lines) and a numerical solution (full lines) for arc-welded aluminium. The numerical solution predicts a more elongated weld pool shape. (b) A comparison between experimental and calculated isotherm shapes in welded aluminium, based on the numerical approach. (c) The thermal cycles. Agreement in all cases appears to be very good. After Kou, S., Simulation of heat flow during the welding of thin plates, *Metall. Transactions,* **12A**, 2025, 1981

heat-flow equations. A comparison between the analytical solution of Rosenthal and a numerical solution of the heat-flow equations is shown in *Figure 1.23*(*a*). A comparison between experimental and calculated isotherms of arc-welded aluminium is given in *Figure 1.23* (*b*), and of the relevant thermal cycles in (*c*); all show very good agreement.

Heat flow in electroslag welding

In electroslag welding the heat source is large and moves slowly. This implies that the material ahead of the moving heat source is preheated to a much higher degree than in normal arc welding so that the heat-flow patterns associated with electroslag welding are not likely to be well described by considering point sources. Hence, the 'quasi-stationary' situation is no longer quite valid. Furthermore, the important role of the slag in this process, with its resistive heating effect, should also be taken into consideration, particularly with regard to its effect on the shape and size of the heat source. In fact, the slag volume is larger in electroslag welding, so that the assumption of a point heat source is not representative of this process.

Attempts to simulate the heat flow of electroslag welding have been made, e.g. by Pugin and Persovskii[14], who expanded the moving point source solution to one of three point sources of different intensities and levels. A more sophisticated approach has recently been made by Liby, Martins and Olson[15]

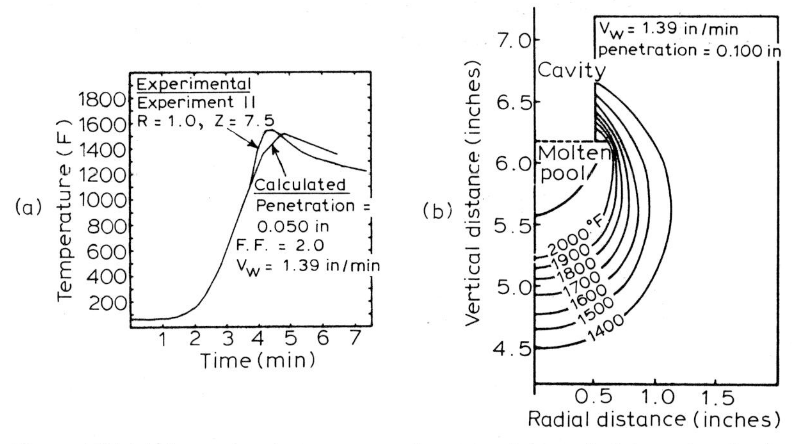

Figure 1.24 (a) Comparison between experimental and theoretical thermal cycles for electroslag welding. (b) Isotherm shapes for the same type of weld. After Liby, Martins and Olson[15]

who used a symmetrical parabolic model to represent the moving molten–solid interface. The fairly satisfactory fit between their model and experimental measurements is illustrated in *Figure 1.24* (*a*), together with typical isotherm shapes as calculated from the model, *Figure 1.24* (*b*).

Weld simulation

It is useful to be able to simulate the weld thermal cycle under laboratory conditions in order to obtain information about microstructural and property changes in the HAZ. Although, in principle, these changes can be observed and measured from real welds, in practice it is more convenient to work with test pieces representative of one, not a range, of microstructures and grain sizes, particularly if mechanical property measurements are required. For this purpose *weld simulators* have been developed and are usually based upon resistance heating and water cooling of samples. Control of the thermal cycle is via a Pt/Pt – 13% Rh thermocouple spot welded to the specimen's surface. A schematic of the equipment is given in *Figure 1.25*. It is possible to program the required thermal cycle to any required temperature–time profile, and to plot this thermal cycle and record phase transformations using a dilatometer. If necessary, a tensile or compressive load can be applied to the sample simultaneous with heating and cooling. Typical operating ranges for commercially available simulators provide a minimum Δt_{8-5} of *ca.* 3 s, giving a maximum heating rate to the peak temperature of about 200 °C s^{-1}. These limitations thus allow most practical weld processes to be simulated, with the exception of TIG or MIG welds of low heat input and other low heat input processes, such as laser- and electron-beam welds. The effect of preheat or reheat is easily accommodated by suitable programming of a computer. The specimen size used is selected to have the typical dimensions of a Charpy V-notch sample. The usual cross-section size (10 × 10 mm) is convenient for microscopic examination, including electron microscopy. The temperature variation at the specimen centre is such that the temperature profile over, e.g.,

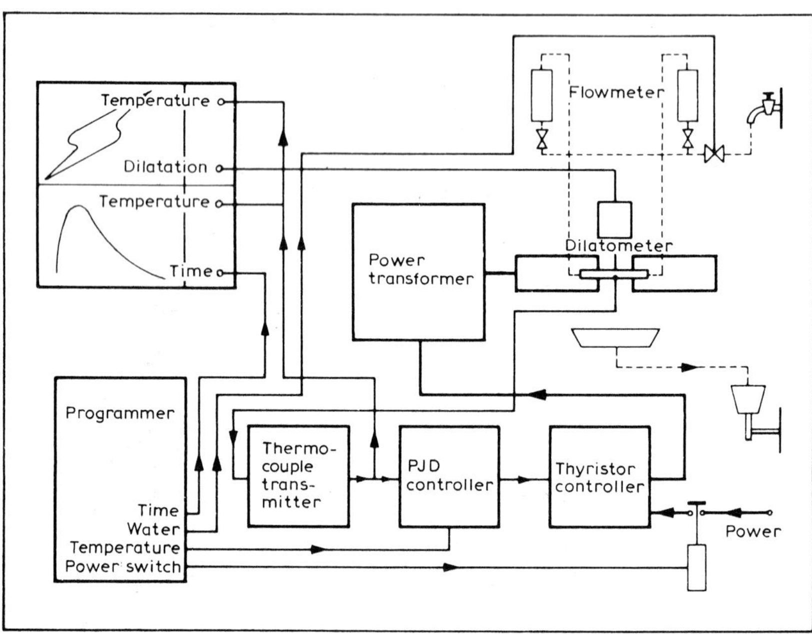

Figure 1.25 Schematic layout of weld simulator equipment

a *ca* 10 mm length, exhibits a parabola along the axis of the sample, with variations of up to *ca* 50 °C at the highest peak temperature. Surface heat losses can also cause variations of up to *ca* 50 °C in sections perpendicular to the axis, depending on the type of metal being studied and the peak temperature used. It is, of course, important to take these variations into account when applying this technique. An example of the temperature variation along the axis of a 6 mm diameter specimen is shown in *Figure 1.26.*

The thermal cycle to be programmed is obtained from heat-flow theory, using, e.g., the Rosenthal or similar equations. It is important to follow this procedure since the final microstructure obtained in the simulated sample depends on the *complete thermal cycle*. It is often stated in the literature that weld simulation is based only on a certain Δt_{8-5}. This is clearly wrong, since the microstructure developed over the temperature range 800—500 °C is dependent upon the grain size developed and the complete thermal history of the HAZ. This is discussed in more detail in Chapter 3.

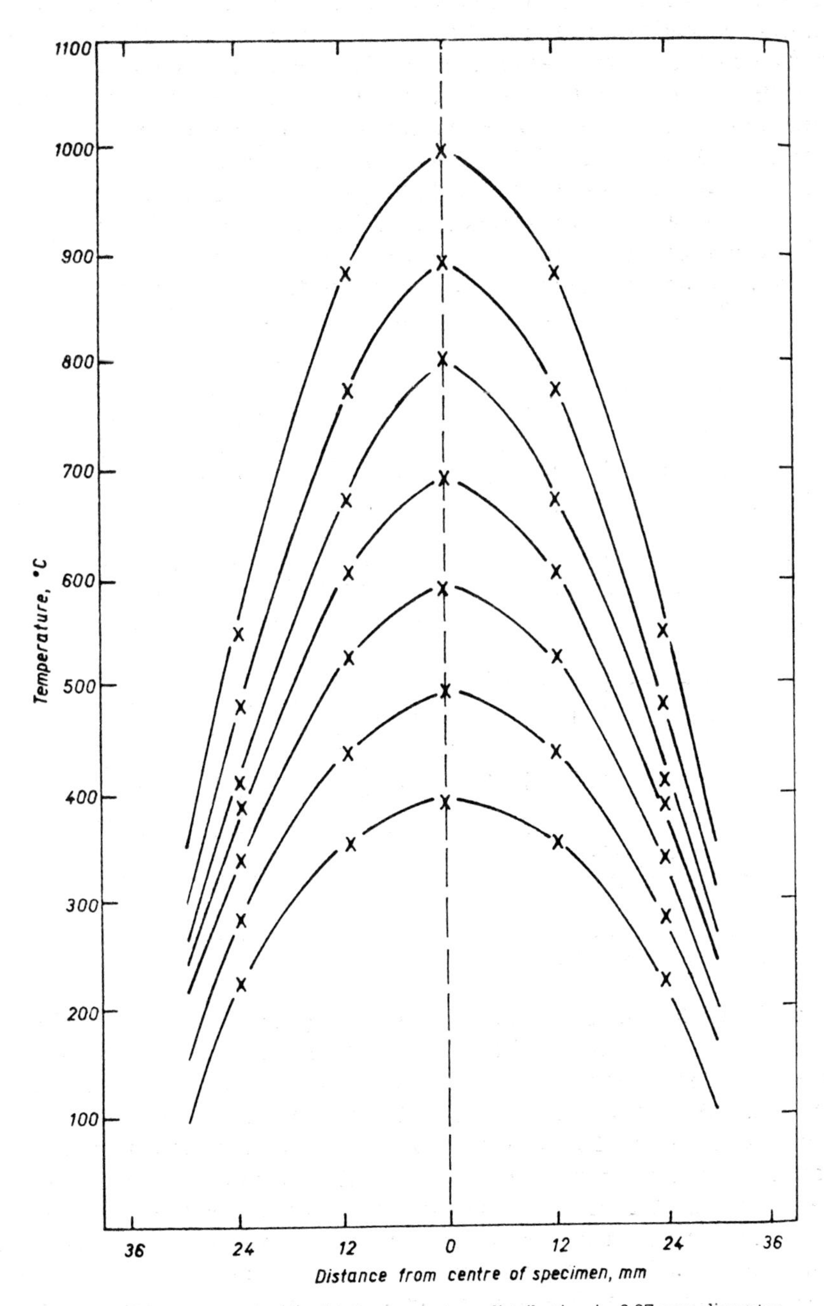

Figure 1.26 Measured longitudinal temperature distribution in 6.37 mm diameter specimens in a weld simulator for temperatures in the range 400–1000 °C. The heated span is 76.2 mm. After Keane, D.M., Bower, E.N. and Hammond, J., Tests during simulation, *Weld Thermal Simulators for Research and Problem Solving*, ed. Dolby, R.E., The Welding Institute, Cambridge, 1972

There have been numerous attempts to compare simulated and real welds. In general, there appears to be fairly satisfactory correlation with respect to both microstructure and mechanical property measurements, although exceptions, particularly in ferritic steels, are often quoted. An investigation by *The Welding Institute*[16], for example, showed that possible reasons for differences in microstructure in these cases could be:

1. Differences in heating rate.
2. Differences in cooling rate.
3. Incorrect measurement of the highest peak temperatures.
4. The large temperature gradient in real welds which can effectively inhibit grain growth.

It is useful to consider these factors in more detail. *The heating rate* is important mainly because of its effect on the rate of *dissolution and coarsening of precipitates* and thus on *grain growth*. Grain growth during heating is also affected by the temperature at which *recrystallization* and/or phase transformation occur. It should also be remembered that recrystallization is, in turn, dependent upon the degree of prior deformation of the base metal. In many welds, as mentioned above, it may not be possible to simulate the fast rate of heating. On the other hand, some high heat input welds, e.g. submerged

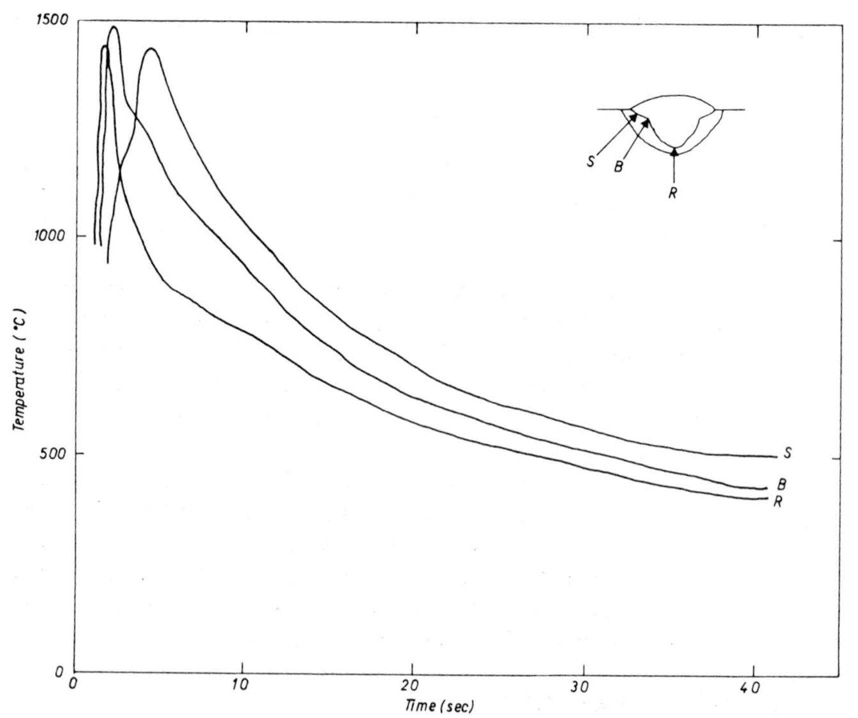

Figure 1.27 Variations in thermal cycles that occur in asymmetrical welds. The thermocouple used in the experiment to give *Figure 1.19* was placed at position *R*. After Kohno, R. and Jones, S.B., *An Initial Study of Arc Energy and Thermal Cycles in the Submerged Arc Welding of Steel*, The Welding Institute, 1978

arc welding of thick plates, may have a slow heating rate, taking several minutes to reach peak temperatures.

The cooling rate, in conjunction with the grain size and prior thermal history, determines the final microstructure. Correct correlation with the Δt_{8-5} of the weld process and geometry used must be employed.

Incorrect measurement of peak temperature can occur, e.g., by using the wrong sort of thermocouple, or by not taking account of the thermal gradient at free surfaces of the test sample, or the gradient along the axis of the sample. This problem is greatest at the highest peak temperatures.

The effect of temperature gradient in real welds is in some ways the most difficult to account for in simulated samples. The problem is that if rapid changes in grain size across the HAZ cause a grain growth inhibiting effect, due to some sort of pinning from the small grains, it is hard to simulate the effect in the type of test samples used for simulation. The effect is clearly worse in low heat input welds, or in high speed welding.

Other problems of simulation concern the effect of the *shape* of real welds on the associated HAZ microstructures. An example of this is shown in *Figure 1.27* for a submerged arc weld. The associated thermal cycles, as measured by thermocouples at the positions marked, are seen to differ somewhat. It appears that if the weld shape is not parabolic, variations in heat flow occur and give rise to different thermal cycles in a given weld. Again, these effects are difficult to simulate and can only be taken into account if the shape of the weld is known beforehand.

Summary of weld thermal cycle results

The Rosenthal equations can usefully describe the arc-welded thermal cycles as experienced by the HAZ of the base metal. Within the limitations of the simplifying assumptions made, there is fairly good correlation between theoretical, real weld and simulated thermal cycles. Refinements to the Rosenthal equations and the use of FEM techniques give a more exact description of the weld thermal cycle, in particular the effect of latent heat release, which is revealed as a distinct kink in the thermal cycle curve. Attempts have been made to model the geometrically larger heat source of the electroslag process using a parabolic form of the heat input instead of assuming a point or line heat source.

The apparently good correlation between predicted and actual thermal cycles is of vital importance to the understanding of the physical metallurgy of welding. It allows reasonable predictions as to the effect of the different welding processes on microstructure and properties. The state of stress that exists in welds during and following the thermal cycle and how this affects microstructure and properties of the weld are now considered.

Residual stresses in welds

Stresses and strains generated by changes in temperature

Consider a butt weld as shown in *Figure 1.28.* A section of material in the vicinity of the weld experiences different rates of expansion and contraction

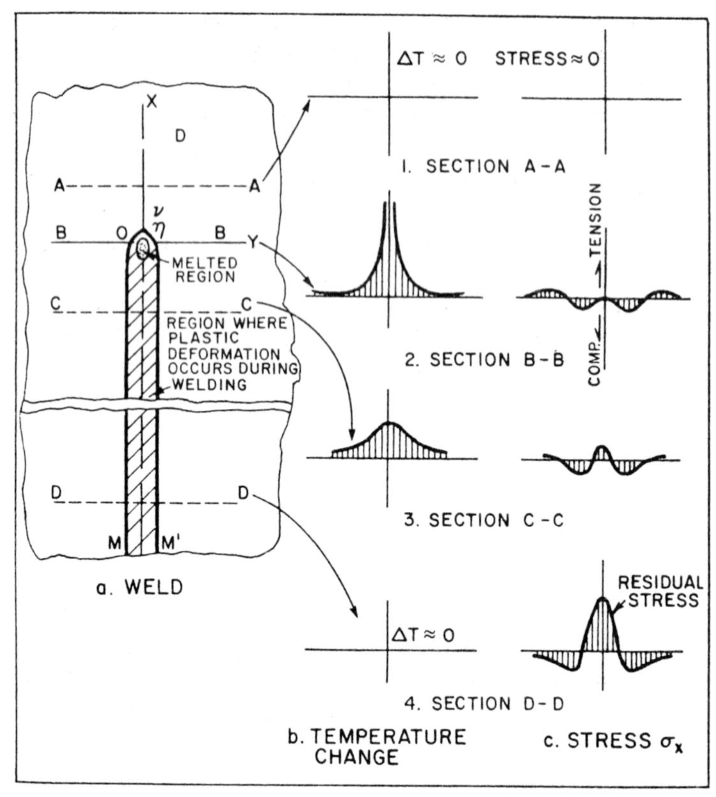

Figure 1.28 Schematic representation of changes of temperature and stresses during welding. After Masubuchi, K., Models of stresses and deformation due to welding — a review, *Modeling of Casting and Welding Processes*, p. 223, eds Brody, H.D. and Apelian, D., Metall. Society of AIME, 1981

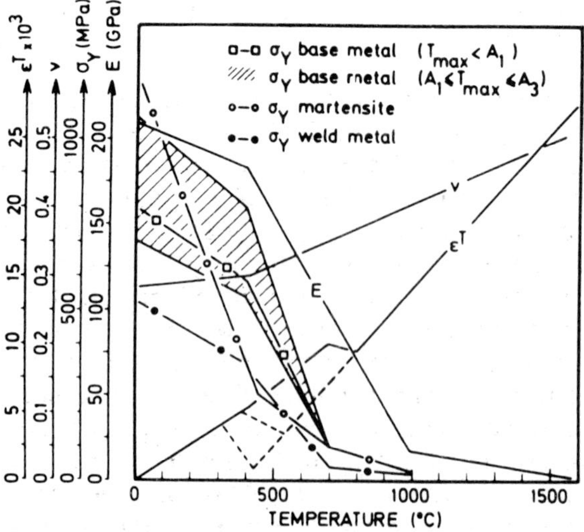

Figure 1.29 Various property changes in a quenched and tempered steel as a function of temperature. After Andersson[13]

compared with other sections around it. In tracing the behaviour of this section during a weld thermal cycle, changes in the section's mechanical properties as a function of temperature, e.g. elastic modulus E, yield strength, Poisson's ratio, etc, have also to be taken into account. These properties can change drastically at temperatures as low as 500—600 °C in steel, as shown in *Figure 1.29.*

Consider qualitatively how this section experiences the weld thermal cycle. As the temperature increases the initial expansion of the section is restrained by material further away from the heat source, which generates compressive stresses in the section, as shown schematically in *Figure 1.30.* The corresponding stress–strain relationship is also shown. The elastic portion of the stress–strain curve is non-linear because of the decrease in E *(Figure 1.29).*

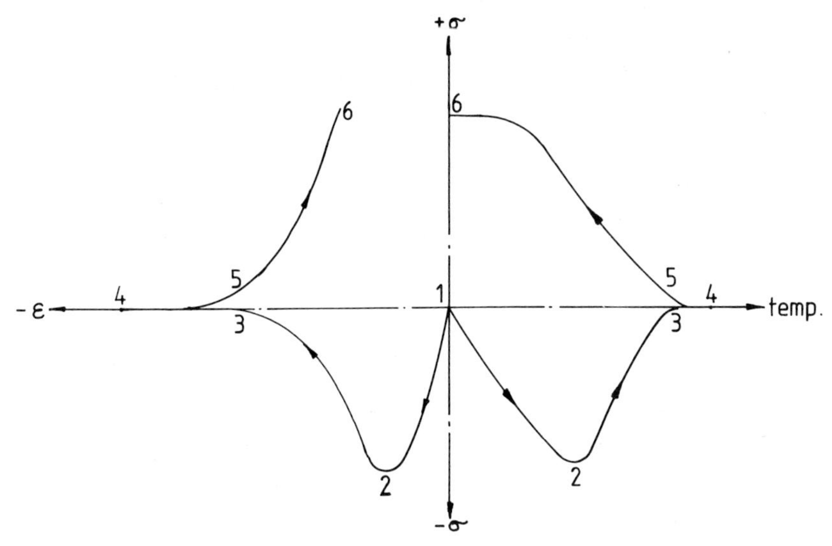

Figure 1.30 Schematic illustration of the variations in stress–temperature and stress–strain during a weld thermal cycle. Point 6 refers to the final residual stress and strain after the element has cooled to ambient temperature

At some critical temperature (point 2), the flow stress of the section is exceeded, and further heating results in a decline in stress as the metal becomes softer. At a temperature of *ca.* 700 °C for a microalloyed steel, the section's stress is practically zero (point 3) and considerable plastic strain may occur. At the peak temperature reached by the section the net strain is represented by the vector 1→4; on cooling, the reverse occurs. As the temperature of the section decreases, resistance by adjacent, hotter material initially reverses the plastic strain pattern until a temperature is reached at which the section's flow resistance increases, which subjects the section to tensile stresses (point 5). Further decrease in temperature is then accommodated by the section in the form of elastic tensile stresses until it cools down completely (point 6). Depending on the rigidity of the structure, the residual stress can (and usually does in thick welded joints) equal the yield stress of the metal. The pattern of residual stresses in the butt weld of *Figure 1.28* thus appears as in *Figure 1.31* (*a*), and the development of these stresses over the whole weld

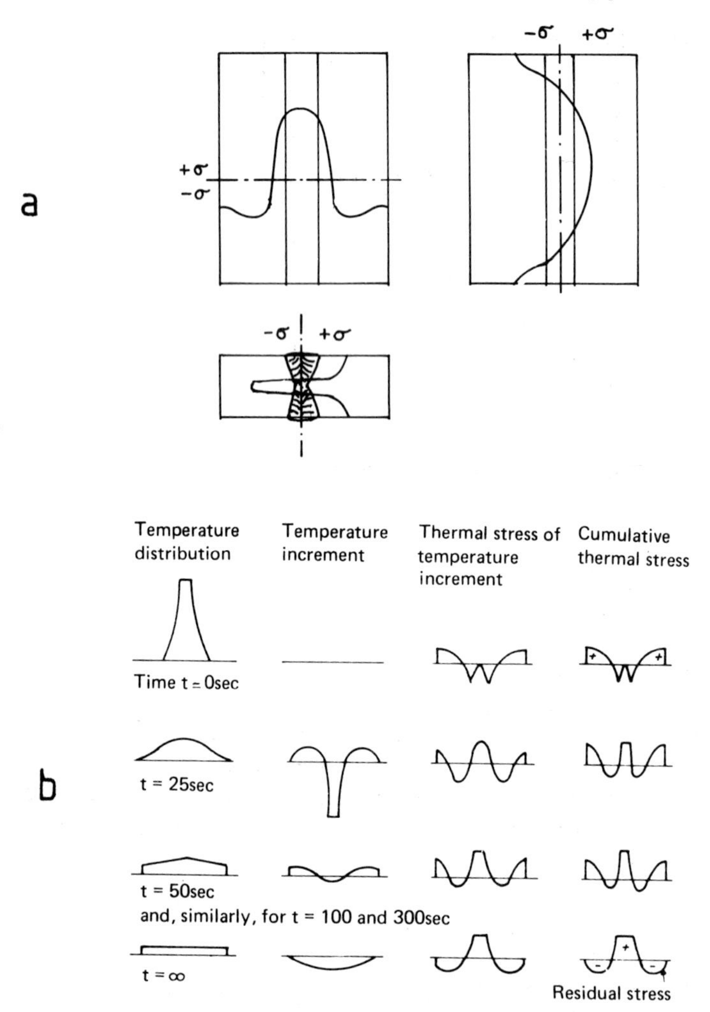

Figure 1.31 (a) Residual stress distributions in a butt-welded plate without fixed ends. (b) The development of residual stresses across a weld as a function of time. After Tall, L., The calculation of residual stresses — in perspective, *Conference on Residual Stresses in Welded Construction and their Effects*, ed. Nichols, R. W., p. 44, The Welding Institute, London, Nov. 1977

cycle is shown in *Figure 1.31* (*b*). The combinations of these residual stresses and the residual strains imposed at high temperature may result in distortion, as illustrated in *Figure 1.32,* although certain welding procedures can reduce this distortion, as indicated in *Figure 1.33.*

Stresses generated by the $\gamma \rightarrow \alpha$ phase transformation

An important contribution to residual stress concerns the dilatation due to phase transformations, e.g., in transformable steel. The temperature at which

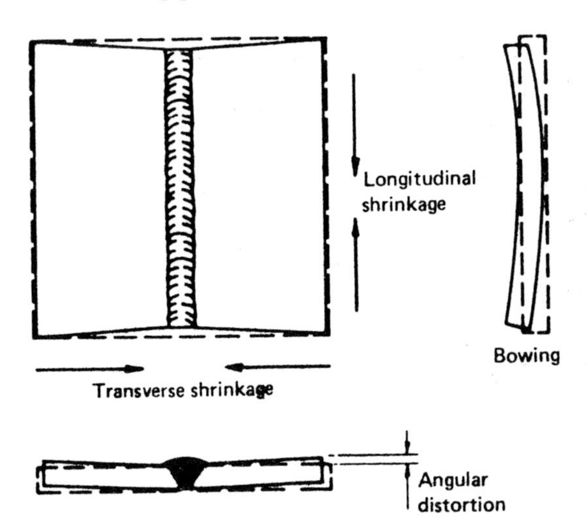

Figure 1.32 Types of distortion in butt welded plates. After Allen, J.S., The effect of residual stresses on distortion, *Residual Stresses and their Effect,* p. 5, ed. Parlane, A.J.A., The Welding Institute, 1981

the phase transformation occurs during cooling of the weld depends on various factors, such as the grain size, the peak temperature reached and the cooling rate (*see* Chapter 2). The magnitude of the volume changes that occur during the transformation can be obtained, e.g., from weld simulation experiments which employ a dilatometer. Examples of such changes in a quenched and tempered steel are shown in *Figure 1.34* as a function of three different peak

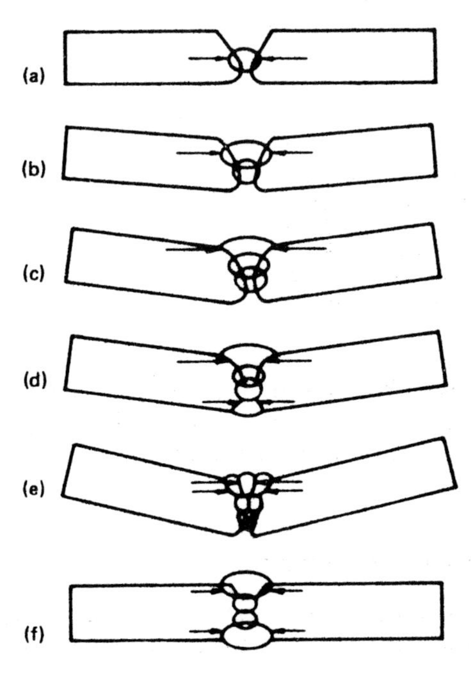

Figure 1.33 Possible welding procedures to avoid distortion of a butt welded plate. After Allen, J.S., The effect of residual stresses on distortion, *Residual Stresses and their Effect,* p. 5, ed. Parlane, A.J.A., The Welding Institute, 1981

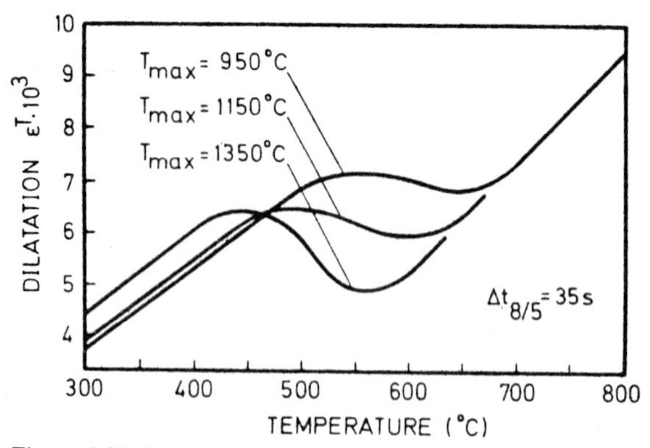

Figure 1.34 Temperature–dilatation curves for a quenched and tempered steel. After Andersson[13]

temperatures. It is observed that the higher the peak temperature, the larger appears to be the dilatation. This effect arises because the lower the $\gamma \rightarrow \alpha$ transformation temperature the larger is the dilatation in transforming from the f.c.c. to the b.c.c. lattice. Larger grain-size material also tends to decrease the transformation temperature because of the reduction in grain boundary surface at which the b.c.c. ferritic phase can nucleate. It thus follows that the

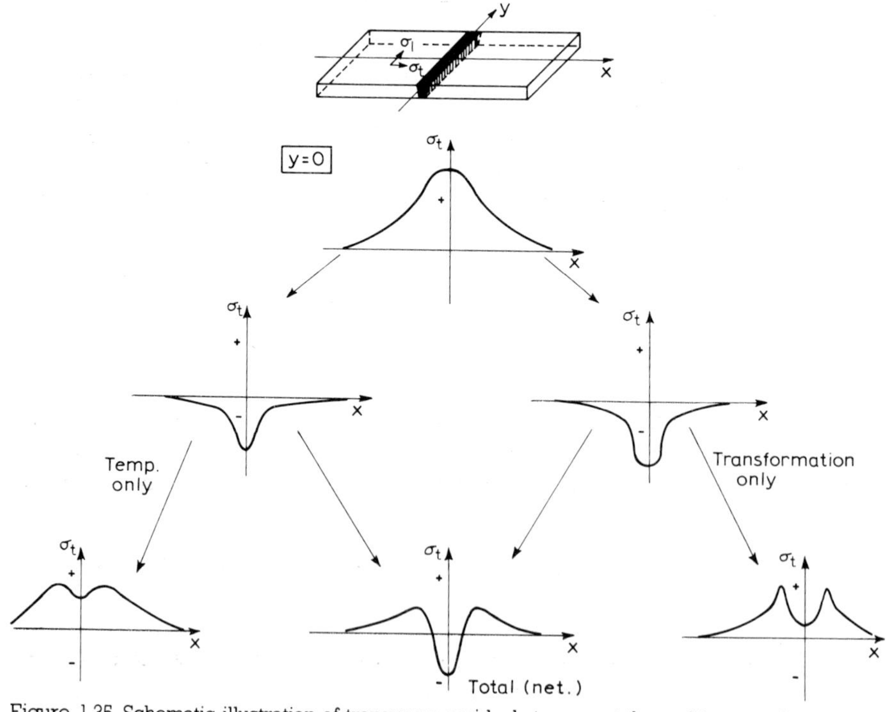

Figure 1.35 Schematic illustration of transverse residual stresses at the weld centre line caused by interaction of shrinkage, quenching and transformation strains. After Macherauch, E. and Wohlfahrt, H., Different sources of residual stress as a result of welding, *Conference on Residual Stresses in Welded Construction and their Effects*, p. 267, ed. Nichols, R.W., The Welding Institute, London, Nov. 1977

largest dilatations – and hence the largest stress changes – are associated with the martensitic $\gamma \rightarrow \alpha'$ transformation, in which case the dilatation is *ca.* 4 %, corresponding to *ca.* 1.4 % linear strain.

The additional stresses due to phase transformation interact with the quenching and shrinkage stresses, as shown schematically in *Figure 1.35*. Note from this figure, however, that the transformation and thermal stresses are not necessarily additive in the sense that the final residual stress is higher.

Measurement of residual stresses in welds

For useful recent reviews of this subject *see*, e.g., Parlane[17] and Procter[18]. The main experimental techniques used are:

1. Mechanical, e.g. hole drilling, or saw cuts.
2. Moiré pattern.
3. X-ray.
4. Ultrasonic.

Of these, the hole drilling and the X-ray approaches appear to be the so-called 'destructive' and 'non-destructive' techniques, respectively, in most common use nowadays.

The *hole drilling technique* is illustrated in *Figure 1.36* with the modified residual stress pattern due to the hole being drilled shown in (*b*). It is seen that when the hole is drilled the strain gauge rosette (*a*) registers a change in strain, so that if the distance between the hole and the gauge length is known, the residual stress, σ_R, can be computed. The relationship between the measured strains in the 0°, 45°, 90° directions, ε_1, ε_2 and ε_3, and the maximum and minimum principal stresses are given by equation (1.28)[18], where the constants

$$\frac{\sigma_{max}}{\sigma_{min}} = -\frac{1}{K_1} \times \frac{E}{2} \times \left\{ \frac{\varepsilon_1 + \varepsilon_3}{1 - \nu K_2/K_1} \pm \frac{1}{1 + \nu K_2/K_1} \left[(\varepsilon_1 - \varepsilon_3)^2 + 2\varepsilon_2 - (\varepsilon_1 + \varepsilon_3)^2 \right]^{1/2} \right\} \qquad (1.28)$$

K_1 and K_2 are calibration factors which depend on the hole size and the distance between the strain gauges and the hole. These constants are given by equation (1.29), were ε_A is the applied strain in the axial direction, ε_A' is the

$$\frac{1}{K_1} = \frac{\varepsilon_A}{\varepsilon_A'} \approx 2.8 \text{ for a 2 mm diameter hole: } \nu K_2/K_1 = -\varepsilon_T'/\varepsilon_A' \approx 0.3 \qquad (1.29)$$

relaxed strain in the axial direction (caused by drilling the hole) and ε_T' is the relaxed strain in the transverse direction.

The direction of the principal stresses φ is given by equation (1.30). The

$$\varphi = \tfrac{1}{2} \tan^{-1} \frac{\varepsilon_1 - 2\varepsilon_2 + \varepsilon_3}{\varepsilon_1 - \varepsilon_3} \qquad (1.30)$$

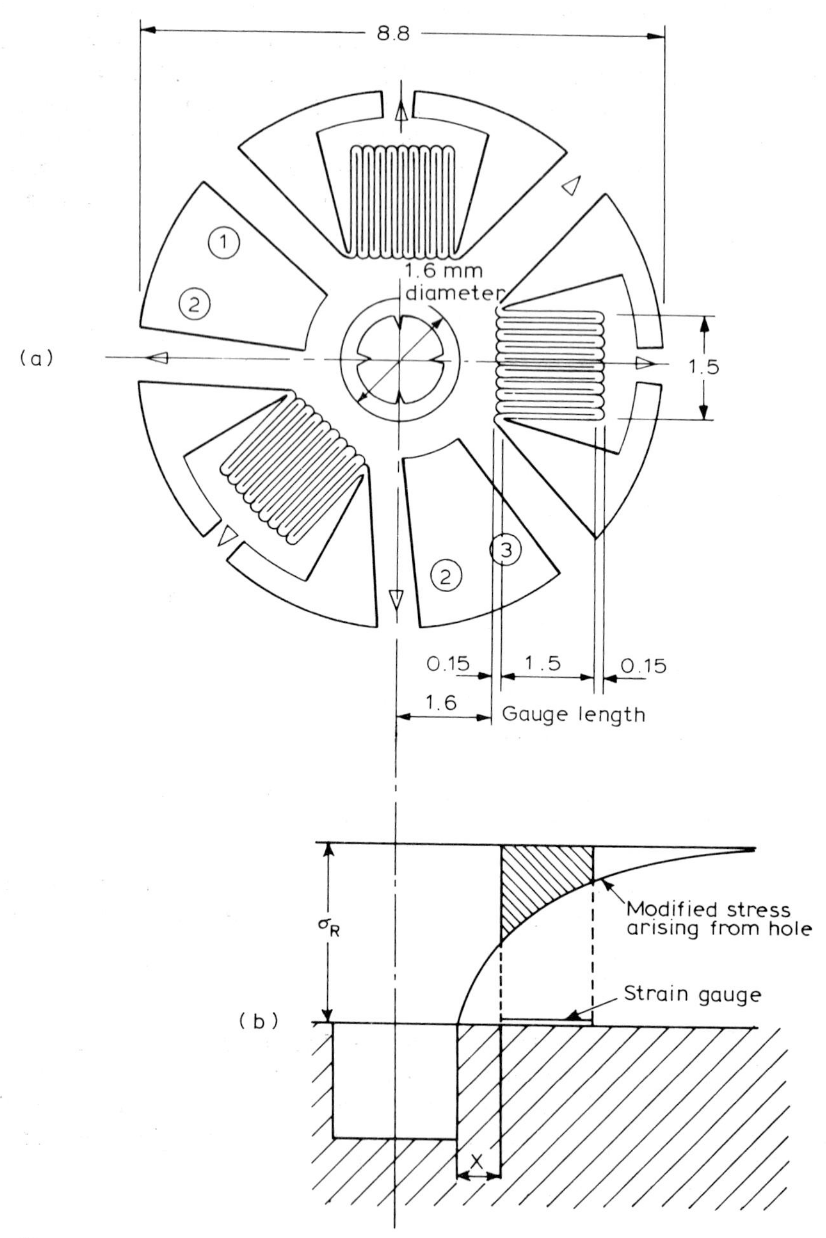

Figure 1.36 Schematic layout of the hole drilling technique in which the change in residual strain due to drilling a hole is measured by the strain gauge rosette, as shown

accuracy of the hole drilling technique is thought to be within 8 — 10 % of the correct value[17].

X-ray diffraction has been used to measure residual stresses for many years, although the technique is considered to be sophisticated and generally requires considerable experience if high accuracy is required. The principle is that elastic deformation of the crystal lattice due to, e.g. residual stresses, affects the lattice spacings and this can be measured using a back-reflection camera (or a suitable goniometer if higher accuracy is needed), as shown in *Figure 1.37*. There are essentially two approaches used: either to compare the stress free and stressed states, or to employ a double exposure technique (or multiple exposure for complicated stress states) using different angles of incidence, ψ, within the plane which encompasses the stress direction, *see Figure 1.37.*

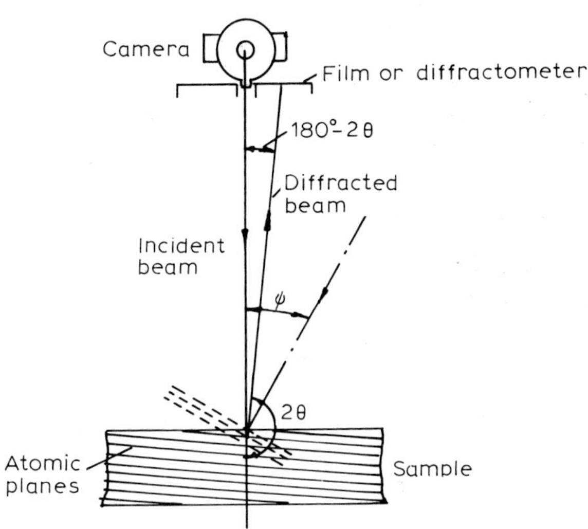

Figure 1.37 Schematic layout of the back-reflection X-ray technique used to measure residual elastic stresses in welds

In the former case, the increase in stress from a reference or unstressed state is given by equation (1.31), where Θ is defined in *Figure 1.37, K* =

$$\sigma = K\,(2\Theta_0 - 2\Theta_1) \tag{1.31}$$

$E/(1+\nu)$, and Θ_0 and Θ_1 refer to the unstressed and stressed states, respectively. The criticism of this technique is that it may be difficult to measure the unstressed state, particularly in welds, so that the double exposure method is often more convenient to use in practice. In this case σ is given by equation

$$\sigma = K\,(2\Theta_{\perp} - 2\Theta_{\psi}) \tag{1.32}$$

(1.32), where $\Theta\perp$ is the angle of diffraction at zero angle of incidence to the principal strain direction, and Θ_{ψ} refers to the angle of incidence, usually 45°. Since K is a function of $(\sin^2\psi)^{-1}$, this technique is often referred to as the $\sin^2\psi$ method, the stress measured being proportional to the slope of the 2Θ *vs* $\sin^2\psi$ curve. On this basis equation (1.32) can be expressed as equation

$$\sigma = K\,\frac{\delta\,(2\Theta)}{\delta\,(\sin^2\psi)} \tag{1.33}$$

(1.33). The advantage of the X-ray method is that it is non-destructive and portable equipment based on a back-reflection camera are commercially available, although their accuracy is at best only within *ca.* 15—20 %[17].

Numerical methods of estimating residual stresses in welds

Despite the accuracy limitations discussed, the most reliable way to estimate residual stresses in welds is by direct measurement. Nevertheless, the development of numerical methods is needed to study the influence of the variables of the process and to be able to make predictions, e.g. concerning whether preheat or postheat is necessary, or concerning susceptibility to hydrogen cracking.

As discussed above, residual stresses in welds result from elastic and plastic deformation and to model this requires complex elastic–plastic treatments. Recent years have seen considerable developments in computer techniques based on the finite-element method for calculating residual stresses even in relatively complicated shapes, and agreement between theoretical prediction and measurements has steadily improved.

In its simplest form, the stress change within a given element as a function of temperature in the vicinity of the weld is given by equation (1.34), where α is

$$\Delta\sigma = \alpha \times \Delta T \times E \tag{1.34}$$

the coefficient of linear expansion and E is the elastic modulus. ΔT here refers to the weld thermal cycle as obtained, e.g., from equation (1.27). In other words, the magnitude of $\Delta\sigma$ depends on three thermal coefficients q, k and λ, and three material properties α, E and σ_y, all of which vary with temperature. In addition, if a phase transformation occurs during cooling it contributes another factor to the equation which, in terms of the dilatational strain, can be expressed as equation (1.35), where $\Delta V/V$ refers to the dilatation, σ is the

$$\varepsilon_d = \frac{5}{6}\frac{\Delta V}{V}\frac{\sigma}{\sigma_y} \tag{1.35}$$

applied stress and σ_y the yield stress of the austenite at the temperature of transformation.

The computational procedure[20] is then to calculate the elastic–plastic strain increments of a network of sectional volumes over a range of time intervals appropriate to the thermal cycle considered. The number of sections and time intervals employed is restricted only by the capacity (and cost) of the computer. Thus, the total strain increment over a given temperature interval may be given by the matrix (1.36), where $d\varepsilon^T$ refers to the thermal expansion

$$\{d\varepsilon^T\} = \{\alpha\}\, dT \tag{1.36}$$

strain. If it is desired to express $d\varepsilon$ in terms of elastic, ε^e, plastic, ε^p, and thermal, ε^T, contributions together, equation (1.37) can be used and, in a

$$\{d\varepsilon\} = \{d\varepsilon^e\} + \{d\varepsilon^p\} + \{d\varepsilon^T\} \tag{1.37}$$

similar way, other material and thermal variables can be accounted for. The increment of stress is then represented by the product of the matrix, $[D]$, and the strain increment, *see* equation (1.38). An example given by Ueda and

$$\{d\sigma\} = [D]\{d\varepsilon\} \tag{1.38}$$

Yamakawa[20] on the application of FEM to calculate stresses of a butt weld is shown in *Figure 1.38* for the cases of (*a*) the change in longitudinal stresses and (*b*) changes in transverse stress. In (*a*) the heat source is marked so that the stress distributions shown are all relative to that position. Thus, the net residual stress distribution is that to the extreme left of the figure. The corresponding

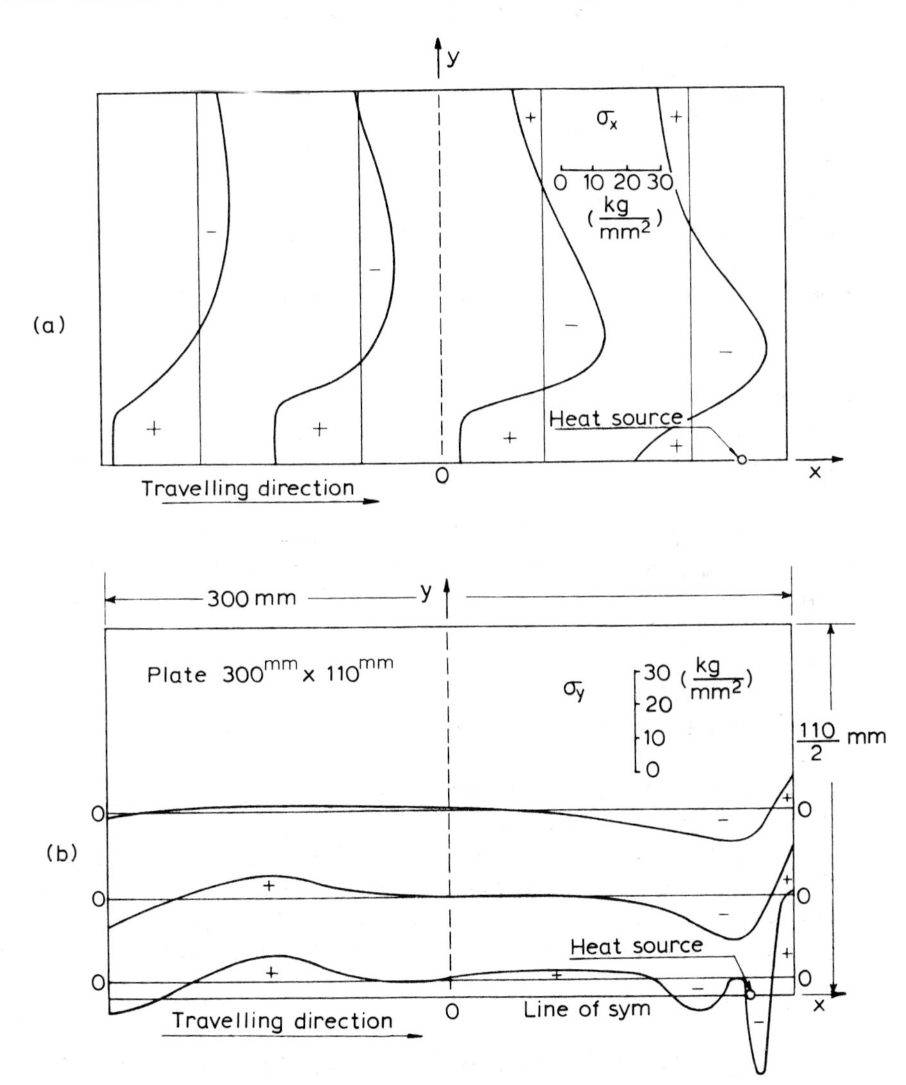

Figure 1.38 FEM-calculated residual stresses in a butt weld which show (a) changes in longitudinal stresses as a function of distance from the arc and (b) changes in transverse stress at various distances from the weld centre line. After Ueda and Yamakawa[20]

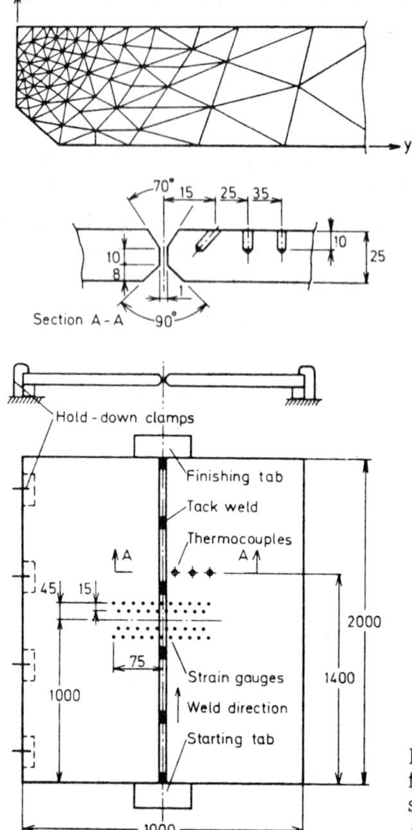

Figure 1.39 An experimental arrangement for the stress analysis of a butt-welded plate, showing the finite element mesh used. After Andersson[13]

property changes which are a function of the temperature of the steel studied, show that both the transient and residual stress reach the yield stress of the metal. The very severe stress associated with the weld at the edge of the plate in the transverse direction, (*b*), is known to produce end-cracking problems in some steels unless preheat is applied.

In a recent, sophisticated application of FEM to a butt-welded fine grained steel Andersson[13] obtained good correlation between calculated and measured residual stresses, as measured by the drilled hole–strain gauge rosette technique. The experimental arrangement of the gauge positions and a detail of the corresponding FEM net is shown in *Figure 1.39*. The results are shown in *Figure 1.40*. The better fit on the upper side of the plate compared to the lower side is thought to be due to the Bauschinger effect. As with the previous results, however, the residual stresses at the weld can be of the order of the flow stress of the steel. These results have important consequences in many welding applications and problems, and some of these are discussed in Chapter 4.

Summary of residual stress work

The severe heating and cooling cycle during fusion welding of thick plates can result in high stresses and strains. These can cause severe distortion or, in cases of rigid restraint, leave residual stresses of the order of the flow stress of the metal. In metals undergoing phase transformations, the residual stress pattern may be further complicated because of the associated dilatational strains.

There are various ways of measuring or estimating residual stresses, based on direct measurement or on numerical techniques. Both approaches are important in practice. The direct measurements may be either destructive (e.g. by monitoring stress relaxation around drilled holes) or non-destructive (e.g. by back-scattered X-ray camera methods). Numerical techniques have been developed to such a degree of sophistication that good correlation between theory and experiment can now be obtained, even on complex weld geometries.

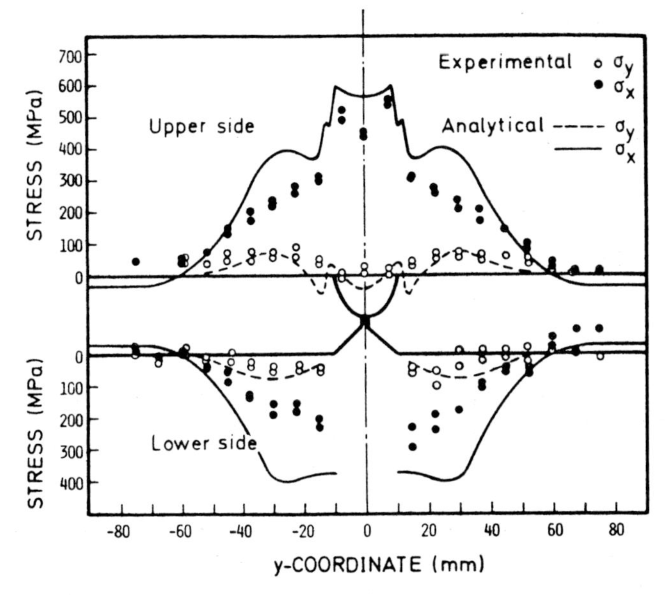

Figure 1.40 Comparison of calculated and measured stress distributions of the arrangement shown in *Figure 1.39*

In this chapter, the fusion welding process has been characterized as a basis for a better understanding of the physical metallurgy of welding. Despite the complexity of the weld thermal cycle, there appears to be a good understanding of the role of the various welding parameters. In particular, understanding the heat flow associated with welding is sufficiently advanced to make useful predictions of the thermal cycle experienced by the plate being welded, and of the stresses associated with the thermal cycle. The metallurgy of the welding process is less well defined and there are certain problems in fully understanding the distribution of gases and elements from fluxes in the weld. Nevertheless, on the whole some useful and constructive deductions can be made as to the likely patterns of behaviour of solidification and development of microstructure in the weld metal and in the heat-affected zone of the base material. This is the subject of the next two chapters.

References

1. Lancaster, J.F., *Metallurgy of Welding,* 3rd edn., George Allen and Unwin, 1980

2. Tylecote, R.F., *The Solid Phase Welding of Metals,* Edward Arnold, 1968

3. Crossland, B., Review of the present state-of-the-art in explosive welding, *Metals Technology,* **3,** 8, 1976

4. Svensson, L-E., ESAB (Gothenburg), personal communication, 1982

5. Almqvist, G., Polgary, C.S., Rosendahl, C.H. and Valland, G., Some basic factors controlling the properties of weld metal, *Proc. Conf. on Welding Research Relating to Power Plant,* p. 204, Central Electricity Generating Board, Leatherhead, 1972

6. Strömberg, J., *Consumables for Shielded Metal Arc Welding,* ESAB Report, Gothenburg, 1967

7. Coe, F.R., *Fluxes and Slags in Arc Welding,* The Welding Institute, Report No. 75/1978/M, Nov. 1978

8. Séférian, D., *The Metallurgy of Welding,* Chapman and Hall, 1962

9. (a) Rosenthal, D., Etude theorique du régime thermique pendant la soudure à l'arc, *Congrés National des Sciences, Comptes Rendus, Bruxelles,* **2,** 1277, 1935; (b) Rosenthal, D., The theory of moving sources of heat and its application to metal treatments, *Transactions ASME,* **68,** 849, 1946

10. Shinoda, T., and Doherty, J., *The Relationship between Arc Welding Parameters and the Weld Bead Geometry,* The Welding Institute, Report No 74/1978/PE, 1978

11. Inagaki, M., and Sekiguchi, H., Continuous cooling transformation diagrams of steels for welding, and their applications, *Transactions of the Japan Institute of Metals,* **2,** No 2, 40, 1960

12. Cacciatore, P., *Modeling of Casting and Welding Processes, (Conference Proceedings)* p.113, eds. Brody, H.A. and Apeliaz, D., Metals Society of AIME, 1981

13. Andersson, B.A.B., Thermal stresses in a submerged-arc welding joint considering phase transformations, *Journal of Engineering Materials and Technology, ASME,* **100,** 356, 1978

14. Pugin, A.I. and Persovskii, G.A., Calculations of the thermal cycle in the HAZ when welding very thick steel by electroslag process, *Automatic Welding,* **16,** No 6, 12, 1963

15. Liby, A.L., Martins, G.P. and Olson, D.L., *Modeling of Casting and Welding Processes (Conference Proceedings),* p.161, eds. Brody, H.D. and Apeliaz, D., Metals Society of AIME, 1981

16. Dolby, R.E. and Widgery, D.J., *Simulation of HAZ Microstructures,* The Welding Institute, April 1972.

17. Parlane, A.J.A., The determination of residual stresses: a review of contemporary measurement techniques, *Conference on Residual Stresses in Welding Construction and their Effects,* p.63, ed. Nichols, R.W., The Welding Institute, London, Nov., 1977

18. Procter, E., Measurement of residual stresses, *Residual Stresses,* The Welding Institute, p. 34, 1981

19. Greenwood, G.W. and Johnson, R.H., The deformation of metals under small stresses during phase transformations, *Proceedings of the Royal Society,* **A283,** 403, 1965

20. Ueda, Y. and Yamakawa, T., Mechanical characteristics of cracking of welded joints, *International Symposium on Cracking and Fracture in Welds,* Japanese Welding Society (Tokyo), p IC 5.1–13, 1971

Further reading

Brody, H.D. and Apelian, D., eds. *Modeling of Casting and Welding Processes,* Metals Society of AIME, 1981

Coe, F.R., *Fluxes and Slags in Arc Welding,* The Welding Institute, Report No 75/1978/M, Nov., 1978

Dolby, R. E., ed., *Weld Thermal Simulators for Research and Problem Solving,* The Welding Institute, Cambridge, 1972

Lancaster, J.F., *Metallurgy of Welding,* 3rd edn., George Allen and Unwin, 1980

Linnert, G.E., *Welding Metallurgy,* 2 Vols., American Welding Society, 1967

Nichols, R.W., ed., *Conference on Residual Stresses in Welded Construction and their Effects,* The Welding Institute, London, Nov., 1977

Residual Stresses and their Effect, The Welding Institute, Cambridge, 1981

Rosenthal, D., The theory of moving sources of heat and its application to metal treatments, *Transactions ASME,* **68,** 849, 1946

Séférian, D., *The Metallurgy of Welding,* Chapman and Hall, 1962

Tylecote, R.F., *The Solid Phase Welding of Materials,* Edward Arnold, 1968

Chapter 2

The weld metal

In this chapter the mechanisms of solidification of the weld metal and the microstructure of the as-solidified weld are considered. Cracking problems associated with the weld deposit are not discussed, since these are dealt with in Chapter 4.

The role of fluxes and slags has been discussed, e.g., in submerged arc and manual metal arc welding, as has their contribution to the final composition of the weld deposit. In addition, it has been noted that if protective gases are used, as in tungsten–inert gas, metal–inert gas or CO_2 gas welding, elements from

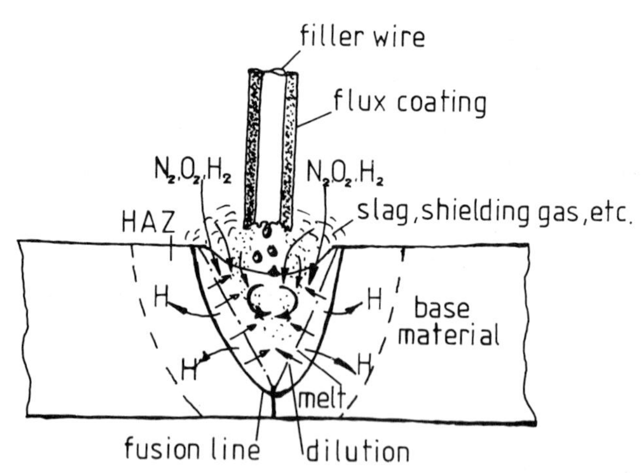

Figure 2.1 Schematic illustration of the various contributions to the composition of a fusion weld

these gases are likely to enter the weld pool. Further alloying additions are obtained from the air and contributed from the base metal being welded, by dilution and even diffusion processes. *Figure 2.1* shows schematically a summary of these various contributions to the final composition of weld deposits.

While it is important to be aware of these various effects, it is not the purpose of this chapter to consider the individual role of each and every alloying element. Instead it is taken for granted that the weld deposit contains a number of alloying additions, many of which are not added deliberately. This chapter concentrates on developing guide lines for understanding the solidification and cooling behaviour of the weld metal; guide lines based on established principles of phase transformations.

Characteristics of weld solidification

The object of this section is to try to apply theories of solidification to the special case of weld solidification, in order to understand the effect of the various welding parameters on the type of microstructure developed. The features of fusion welding which have to be taken into consideration are:

1. The weld pool contains impurities.
2. Dilution.
3. Considerable turbulence — therefore good mixing occurs in the molten metal.
4. The molten metal volume is small compared with the size of the mould (the base metal).
5. The composition of the molten metal and mould are very similar.
6. There are large temperature gradients across the melt.
7. Since the heat source moves, weld solidification is essentially a dynamic process; the solidification behaviour is thus dependent on the welding speed.
8. In high energy welds (e.g. electroslag) or multi-run welds in which the base metal is preheated, temperature gradients and hence solidification behaviour are affected.

In many respects, weld solidification is a fundamentally different process to that of *ingot casting;* on the other hand it has certain features in common with *continuous casting*.

Geometry of the weld melt

As discussed in Chapter 1, equation (1.12) can be used to estimate the shape of the melt as a function of different materials, welding speeds, etc. A convenient way of comparing the *shape* of melts, although not necessarily their physical size, is by plotting the isotherms for different processes. From equation (1.14), the distance between the isotherms, $i_{x,\ y,\ z}$, is a proportionality which depends on the direction from the heat source, $r_{x,\ y,\ z}$, given by equation (2.1). Here, q is the heat flux, λ is the thermal conductivity, d is the thickness if

$$i \propto q/(\lambda v d) \qquad (2.1)$$

a thin plate is considered, and v is the velocity of the heat source. Using this approach, a comparison between the isotherm distributions can be estimated for different materials, as shown in *Figure 2.2.* As a basis of comparison all the isotherms in this and the following figures are plotted in the x–y plane of the plate being welded. The significant difference in λ between, e.g., Al and

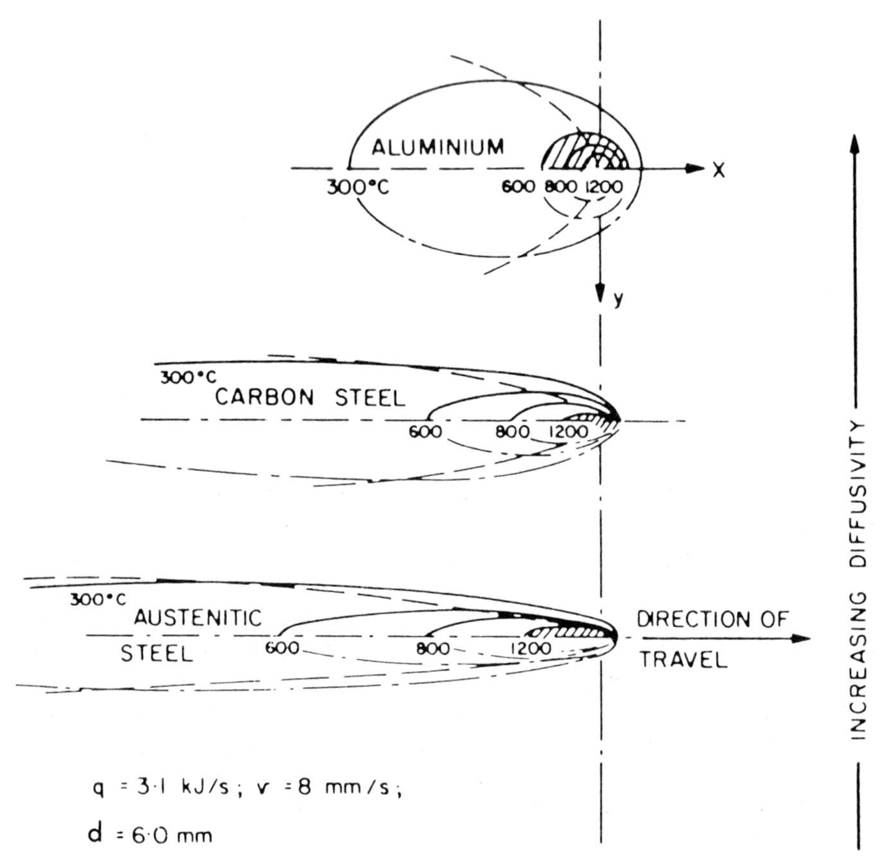

Figure 2.2 Effect of changes in alloy and thermal conductivity on isotherm shapes as calculated from the Rosenthal equations. After Gray, T.G. *et al.*, *Rational Welding Design*, Newnes-Butterworths, 1975

austenitic steel (*see Table 1.5*) is very apparent in *Figure 2.2. Figure 2.3* compares the geometries as a function of welding speed for a carbon steel. As would be expected, the higher welding speed predicts a much narrower weld deposit. The effect of an increase in the plate thickness being welded is illustrated in *Figure 2.4.* In this case the much higher efficiency of cooling of thick plates has the effect of reducing the size of the weld deposit. Obviously, the use of these equations gives only a semi-quantitative idea of the size of the weld melt, but it is a useful way to compare the effect of the various welding parameters. As is discussed below the mechanism of solidification is also affected by these changes in geometry.

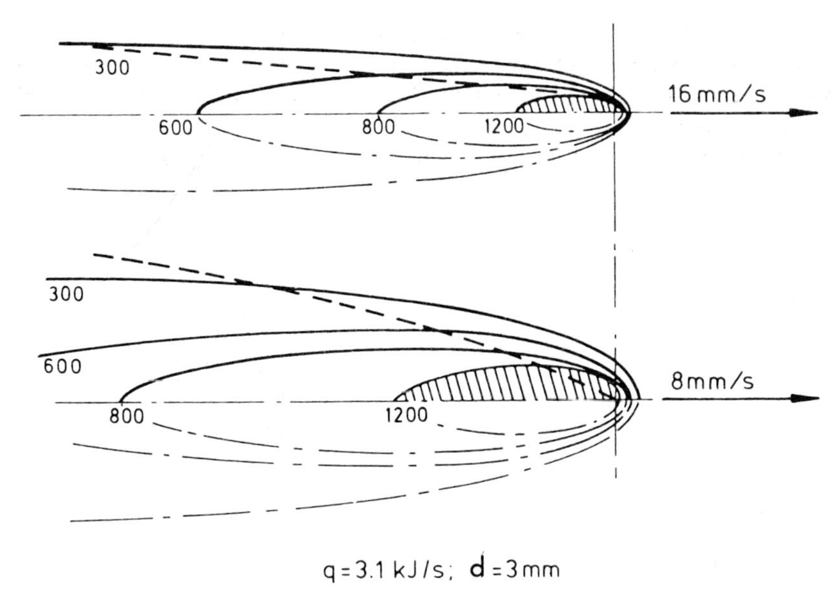

Figure 2.3 Influence of welding speed on isotherm shape in a mild steel. After Gray, T.G. *et al.*, *Rational Welding Design*, Newnes-Butterworths, 1975

A more detailed, quantitative idea of the shape of the melt can be obtained using the FEM technique to solve equation (1.27) (as discussed in Chapter 1) and an interesting example of the weld geometry using this approach is illustrated in *Figure 2.5* for a bead-on-plate weld of aluminium. In this work an attempt to include the effect of a finite volume of the liquid phase was made, and *Figure 2.5* shows the predicted shape of the solidification range as bounded by the liquidus and solidus isotherms (shaded), and projected in three dimensions. The shaded zone, indicating the solidification range of the aluminium alloy in this figure, is usually referred to as 'the mushy zone' because of its mixed solid–liquid character. The *mushy zone* is discussed more specifically in terms of alloy composition in the sections that deal with the mechanisms of solidification.

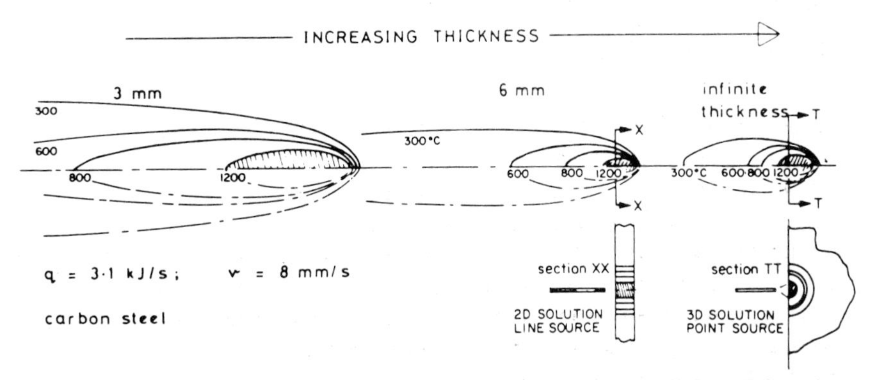

Figure 2.4 Effect of changes in steel plate thickness on isotherm shape. After Gray, T.G. *et al.*, *Rational Welding Design*, Newnes-Butterworths, 1975

Interestingly, if the geometry of the weld pool is to retain a constant shape, then certain requirements concerning the speed of crystal growth relative to the speed of welding have to be fulfilled. This can be understood by considering the liquidus isotherm in *Figure 2.6*. Assume, to begin with, that crystal growth always occurs in a direction perpendicular to the isotherms. If the velocity of welding is represented by a vector $\boldsymbol{v}$, and the speed of crystal growth by a vector

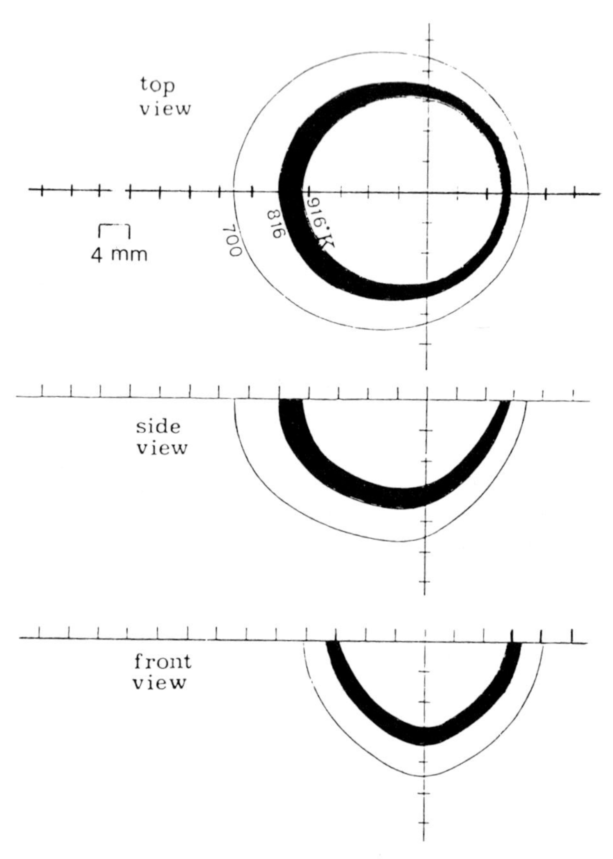

Figure 2.5 The temperature distribution in a bead-on-plate weld of aluminium; v = 20 mm s^{21}, heat input = 430 W. After Kou, S., Simulation of heat flow during the welding of thin plates, *Modeling of Casting and Welding Processes,* p. 129, eds Brody, H.D. and Apelian, D., Metals Society of AIME, 1981

$\boldsymbol{R}$, then condition (2.2) must be met, where Θ is the angle between $\boldsymbol{R}$ and $\boldsymbol{v}$ and,

$$\boldsymbol{R} = \boldsymbol{v} \cos \Theta \qquad (2.2)$$

thus, effectively defines the position on the liquidus isotherm. Since the welding speed, $\boldsymbol{v}$, is constant, $\boldsymbol{R}$ has to vary considerably depending on the position at the liquidus. It thus follows that crystals growing at the weld centre line behind the moving heat source ($\Theta \approx 0°$) grow fastest, while crystals at the

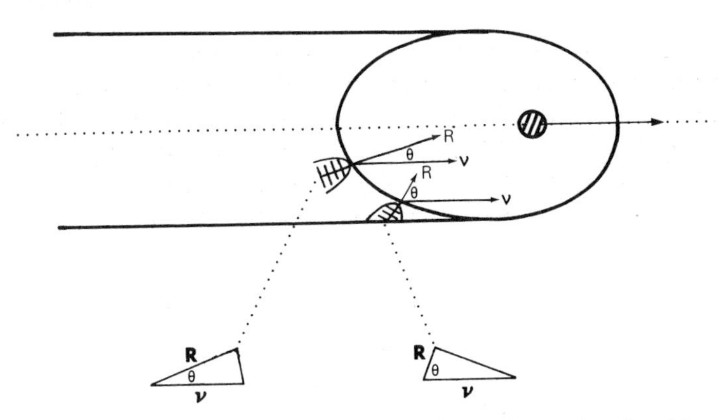

Figure 2.6 The relationship between crystal growth speed and welding speed in terms of rate vectors

edge of the weld ($\Theta \approx 90°$) grow slowest. This effect has important consequences on the microstructure and toughness of welds. Consider, for example, two welds of different speeds, *Figure 2.7* (*a*) and (*b*). In (*a*) crystal growth at low welding speed is able to keep up with the moving heat source in an orderly and symmetrical way, as shown. If the welding speed is increased, however, the shape of the liquidus isotherm (*b*) is altered (*see*, e.g., *Figure 2.3*), with the result that growing crystals have to make sudden changes in direction

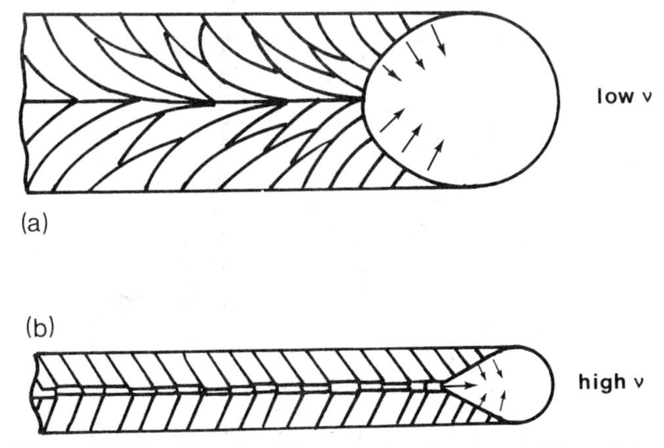

Figure 2.7 The effect of welding speed on the shape of the melt and on the mode of crystal growth. Compare with *Figure 2.3*

at the weld centre line in order to maintain continuity of the weld. Abrupt changes in growth direction, such as shown in *Figure 2.7* (*b*) can be detrimental to the toughness of welds due to increased risk of segregation at the centre line. This problem is discussed later. A micrograph of a weld that illustrates the crystal growth characteristics featured in *Figure 2.7* (*b*) is shown in *Figure 2.8*.

It is assumed in equation (2.2) that crystal growth is isotropic and can always occur along the steepest temperature gradients of the weld. In practice, crystal growth tends to grow along certain preferred crystallographic

Figure 2.8 Crystal growth direction changes in a steel weld. Both abrupt and gradual changes in direction are observed (*see* arrows). Courtesy of ESAB, Gothenburg (× 15; then reduced by one-fifth in printing)

directions, and this is <100> in the case of cubic crystals such as b.c.c. and f.c.c. metals. The preferential growth in the <100> direction is thought to be because this direction is the least close packed; this implies that as crystallization occurs from the random atomic arrangement in the liquid, the looser packing of atoms in the <100> direction effectively allows a faster growth speed than in other (closer packed) directions. In order to take this additional, crystallographic feature into account, equation (2.2) has to be modified. If Θ^d is defined as the angle between the appropriate <100> growth direction and the direction *closest* to the steepest temperature gradient (*see Figure 2.9*), then the actual crystal growth velocity, $\boldsymbol{R}'$, is given by equation

$$\boldsymbol{R}' = \boldsymbol{R} \cos \Theta^d \tag{2.3}$$

(2.3). Hence, the fastest growth speeds occur when the <100> direction coincides with the steepest temperature gradient, i.e. when $\Theta^d = 0$.

Figures 2.7 and *2.8* show that it is often necessary for crystals to change orientation during growth in order that they may continue to follow the

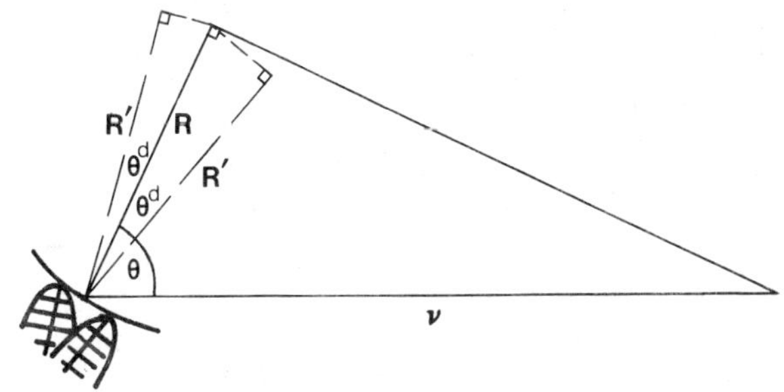

Figure 2.9 Relationship between crystal growth normal to the maximum temperature gradient ($\boldsymbol{R}$), and along the appropriate <100> direction ($\boldsymbol{R}'$)

steepest temperature gradient of the weld pool, and this occurs irrespective of welding speed. This implies that crystals growing in a certain <100> direction are unable to maintain this direction, because as the heat source is moving away the direction of the steepest temperature gradient is effectively changing. In terms of equation (2.3), Θ^d approaches Θ, so that $\boldsymbol{R}' \rightarrow 0$. In order to maintain growth continuity, the solidifying crystals have to change orientation and this

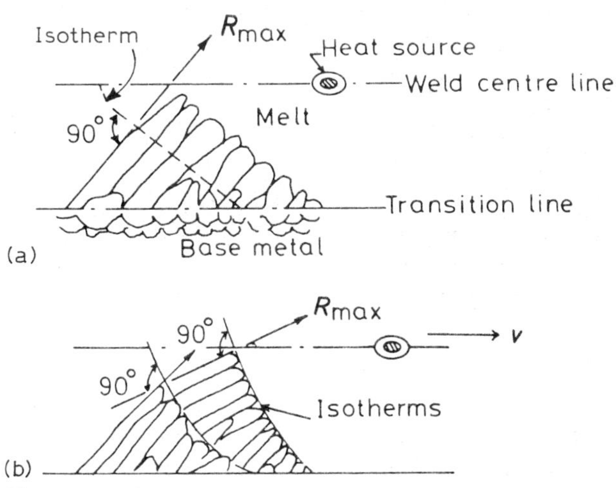

Figure 2.10 Crystal growth occurs along <100> directions that correspond most closely to the steepest temperature gradients in the weld pool. Because of the movement of the arc this may necessitate the re-nucleation of crystals from one <100> direction to another, as shown in (b)

requires them to *re-nucleate* along another more appropriate <100> direction. *Figure 2.10* illustrates this process. However, the mechanism by which changes in growth direction occurs is not properly understood, although it seems likely that re-nucleation at an atomic level on the surface of the existing crystals can occur without difficulty (*see* next section). The possibility that re-nucleation occurs by broken-off pieces of dendrite re-adhering at the solid–liquid interface seems unlikely, although some researchers have suggested this[1]. The turbulence in the weld pool may indeed tend to break-up dendrites, but it seems likely that the same turbulence may well make it difficult for the dendrite particles to re-adhere to the solid–liquid interface. This problem, together with other detailed aspects of the mechanism of solidification in welds is considered in the next section.

Epitaxial solidification

Irrespective of the type of fusion welding process considered, some dilution always occurs. In fact, it is extremely important that dilution does occur since this is the only way to ensure that the surface of the base metal in contact with the liquid weld metal is *clean*, i.e. free of oxide films or other impurities.

The way in which the liquid weld metal solidifies is now discussed. The various boundary conditions concerning the geometry and temperature distributions remain as defined at the beginning of this chapter. The fact that dilution occurs implies that the base metal is locally heated to its melting point and this has certain repercussions concerning the microstructure of the base metal. The most important of these is that *grain growth* is likely to occur in the

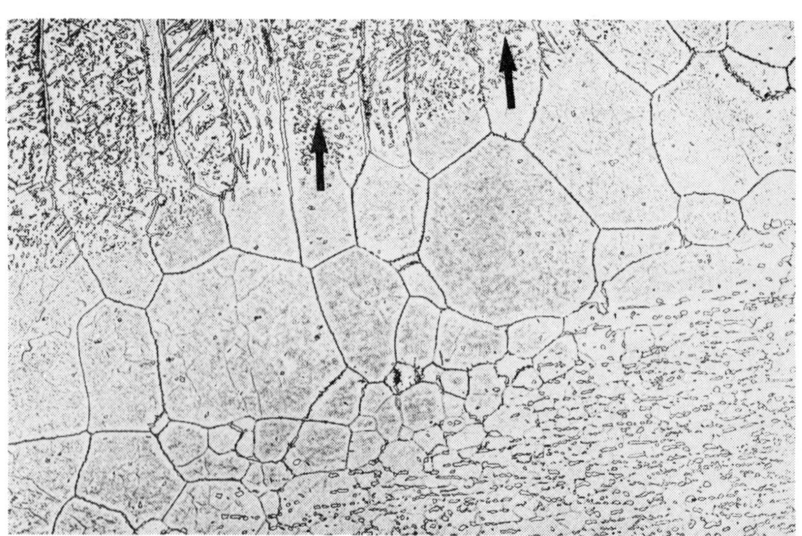

Figure 2.11 The columnar grains of the weld deposit are formed by epitaxial solidification from the base metal grains at the fusion line, as shown in this micrograph of a MIG-welded stainless steel. After Honeycombe, J. and Gooch, T.G., *Weldability of Stress Corrosion Resistant Stainless Steel: Interim Report,* The Welding Institute, Report No 13/1976 1M, 1976 (×100; then reduced by three-tenths in printing)

base metal, and this will be greatest at the highest temperatures, i.e. nearest the *fusion line* between the melted and unmelted metal. The reason that grain growth is important in this respect is that it is well established that, in fusion welding, initial solidification occurs *epitaxially*. That this is so can be easily seen from the micrograph in *Figure 2.11*, which shows that the crystals of the weld metal have clearly derived from grains of the base metal. This result has also been confirmed by X-ray diffraction[2].

Epitaxial solidification is a *heterogeneous* process and is illustrated in *Figure 2.12*. A solid embryo of the weld metal forms at the melted-back surface of the base metal. The shape of the embryo depends on the various surface

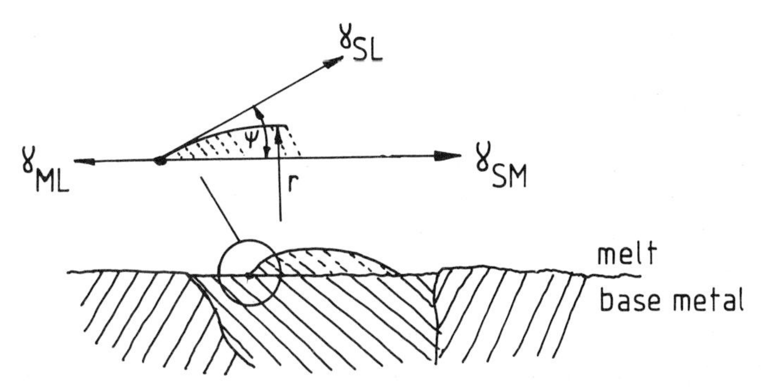

Figure 2.12 Heterogenous nucleation of a hemispherical cap of solid nucleus at the fusion boundary

energies of the system, γ_{ML} (the base metal–liquid surface energy), γ_{SL} (the solidified weld metal–liquid surface energy) and γ_{SM} (the surface energy between the solid weld metal and the base metal). Assuming γ_{SL} is isotropic, it can be shown that, for a given volume of the embryo, the interfacial energy of the whole system can be minimized if the embryo has the shape of a spherical cap. The angle that this cap adopts relative to the original interface is ψ, and this is defined as the 'wetting' angle. The wetting angle is thus controlled by the balance of forces generated by the respective surface energies, as described by equations (2.4) or (2.5). Since initial solidification of the weld metal occurs

$$\gamma_{ML} = \gamma_{SM} + \gamma_{SL} \cos \psi \tag{2.4}$$

$$\cos \psi = \frac{(\gamma_{ML} - \gamma_{SM})}{\gamma_{SL}} \tag{2.5}$$

epitaxially and the weld metal is usually very similar in composition to the base metal, the approximations (2.6) are valid, so the wetting angle $\psi \approx 0$. Hence,

$$\gamma_{SM} \approx 0 \text{ and } \gamma_{ML} \approx \gamma_{SL} \tag{2.6}$$

there is a negligible barrier to solidification of the weld metal, i.e. no undercooling of the melt is needed, and solidification occurs uniformly over the whole grain of the base metal. Of course, at an atomic level, even the initial stage of crystal growth by epitaxial solidification brings about some increase in surface energy due to the atom-high ledges which bound the new growing interface (*see* e.g. ref. 3), but to a first approximation this can be neglected.

Returning briefly to the problem of changes in crystal growth direction because of the moving heat source, this is not such a low energy process as epitaxial solidification because of the necessity to create a new grain boundary. However, since a high angle grain boundary has about a third of the surface energy of the metal, the total energy for nucleation in this case is still relatively low. In fact, from equation (2.5), $\psi \approx \cos^{-1}(2/3)$.

Interestingly, the initial solidification of weld metal is very different from ingot casting solidification in this respect. It is, perhaps, useful at this stage to consider this difference in a little more detail. Since ingot casting involves a substrate material essentially different from the cast metal, the wetting angle is likely to be relatively large. However, it is known from experiment that ingot casting solidification is also a heterogeneous process, usually initiated at discontinuities in the mould surface. On this basis it can be shown that the total free energy of solidification, ΔG, is given by expression (2.7), where r refers to

$$\Delta G = \{-\frac{4}{3}\pi r^3 \Delta G_v + 4\pi r^2 \gamma_{SL}\} S(\psi) \quad (2.7)$$

the radius of the cap (*Figure 2.12*), ΔG_v is the volume free energy of the embryo, and $S(\psi)$ is a shape factor which depends on the wetting angle, and is given by equation (2.8). $S(\psi)$ has a numerical value ≤ 1. By differentiating

$$S(\psi) = \frac{(2 + \cos\psi)(1 - \cos\psi)^2}{4} \quad (2.8)$$

equation (2.7) and equating to zero, it can be shown that the critical size of the embryo, r^*, and its critical free energy, ΔG^*, prior to rapid growth are given by equations (2.9) and (2.10). A physical understanding of these equations can be

$$r^* = \frac{2\gamma_{SL}}{\Delta G_v} \quad (2.9)$$

$$\Delta G^* = \frac{16\pi\gamma_{SL}^3}{3\Delta G_v^2} \times S(\psi) \quad (2.10)$$

obtained by considering that ΔG_v refers to a certain number of atoms which cluster together from the liquid to form a small crystalline embryo at the mould surface. If the liquid temperature is above the melting point, the embryo is unstable. However, as the temperature decreases the clusters of atoms become more stable and grow until they reach such a size that very rapid growth of the solid can occur. This happens when the total volume energy of the embryo plus its interfacial energy with the mould (γ_{SM}) is less than the energy of the liquid in contact with the mould. The relative interdependence of γ_{SL}, ΔG_v and r is thus shown by equation (2.9). Note, however, how extremely sensitive ΔG^* is to γ_{SL} and ΔG_v. In weld solidification, $S(\psi) \approx 0$, so that solidification has no barrier (due to γ_{SL}) and occurs spontaneously below the melting point of the base

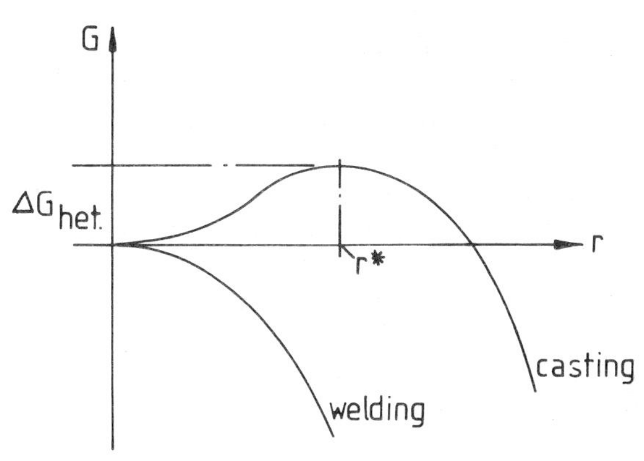

Figure 2.13 The different free energy driving forces for casting and welding solidification. Both are heterogeneous processes, but epitaxial solidification in welding reduces the nucleation barrier to virtually zero

metal. A comparison between weld and casting solidification is shown in *Figure 2.13,* based on equations (2.7)—(2.10).

The conclusion is that the initial solidification of a weld is by epitaxial growth from the grains of the base metal. Hence, the initial crystal size of the weld metal is inherited directly from the grain growth zone of the base metal. This is a problem in high energy processes, such as submerged arc and electroslag welding, where grain growth of the base metal can be considerable. The inherited crystal structure of the weld metal is thus correspondingly coarse.

Crystal growth and segregation

It has already been established that as the weld metal continues to solidify the grains that have a <100> axis approximately parallel to the steepest temperature gradient in the liquid quickly outgrow those grains with less favourable orientation. The growth of a favourably oriented crystal under the conditions expected of fusion welding, i.e. an initially high temperature gradient and good mixing in the liquid is now discussed. For the sake of simplicity the solidification of a binary alloy is considered first, followed by a discussion of the possible consequences of the model with regard to more complex alloying.

Consider the solidification of alloy X_0 in the binary system shown in *Figure 2.14*. In solidification theory, it is normal to discuss three limiting cases, any of which may govern crystal growth.

1. Solidification occurs under complete equilibrium conditions.
2. Solidification is not an equilibrium process and no diffusion occurs in the solid, but perfect mixing occurs in the liquid.
3. Solidification is not an equilibrium process, no diffusion occurs in the solid, and mixing in the liquid is not perfect, but occurs by diffusional processes only.

The rapid rates of cooling of fusion welds imply that weld solidification cannot occur under equilibrium conditions, so point 1 can be ruled out. To begin with, the assumption that negligible diffusion in the solid phase is a good first approximation is made. In the initial stages of solidification good mixing of the liquid seems likely due to the turbulence of the melt. However, as the heat source moves away and solidification goes to completion it seems likely that

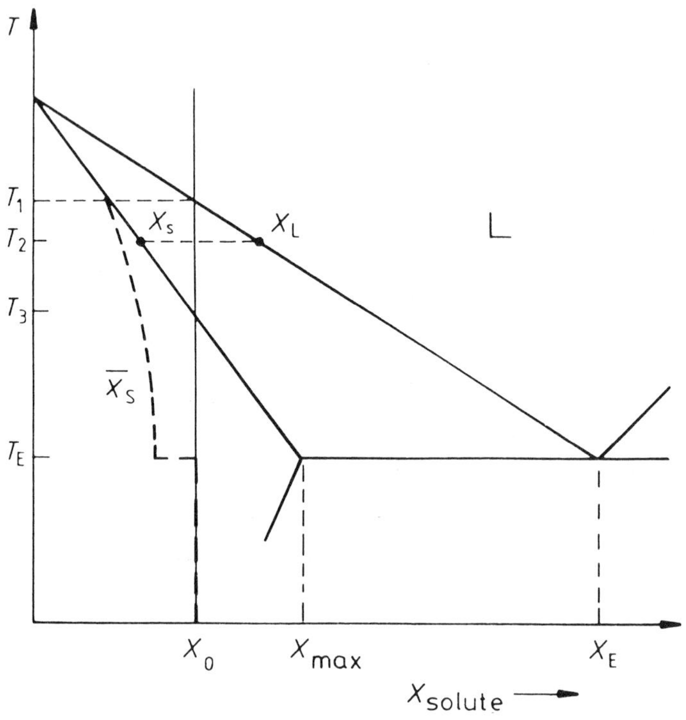

Figure 2.14 The non-equilibrium solidification of an alloy, X_0, in a simple binary system

convectional mixing gives way to diffusional mixing. On this basis it may be conjectured that weld solidification lies somewhere between cases 2 and 3, or, more accurately, progresses on a path from 2 to 3. Both these cases are now considered in detail.

No diffusion in solid; perfect mixing in liquid

Consider a cylindrical volume element of the growing crystal, *Figure 2.15 (a)*. Referring to *Figure 2.14*, an alloy of composition X_0 has a liquidus temperature of T_1. However, since solidification of the alloy occurs over a finite temperature range (T_1 to T_3), partitioning has to take place. It is useful to define a *partitioning coefficient*, k, by equation (2.11), where X_S and X_L are the

$$k = \frac{X_S}{X_L} \tag{2.11}$$

mole fractions of solute in the solid and liquid which coexist with one another at a given temperature. Thus, at a temperature T_1, the first solid to form contains kX_o of solute. Since $kX_o < X_o$, this initial solid is *purer* than the liquid from which it forms. This implies that solute has to be rejected into the liquid, raising its concentration to above X_o. The temperature of the interface must therefore decrease below T_1 before further solidification can occur and the next layer of solid will be richer in solute that the first. As this sequence of events continues

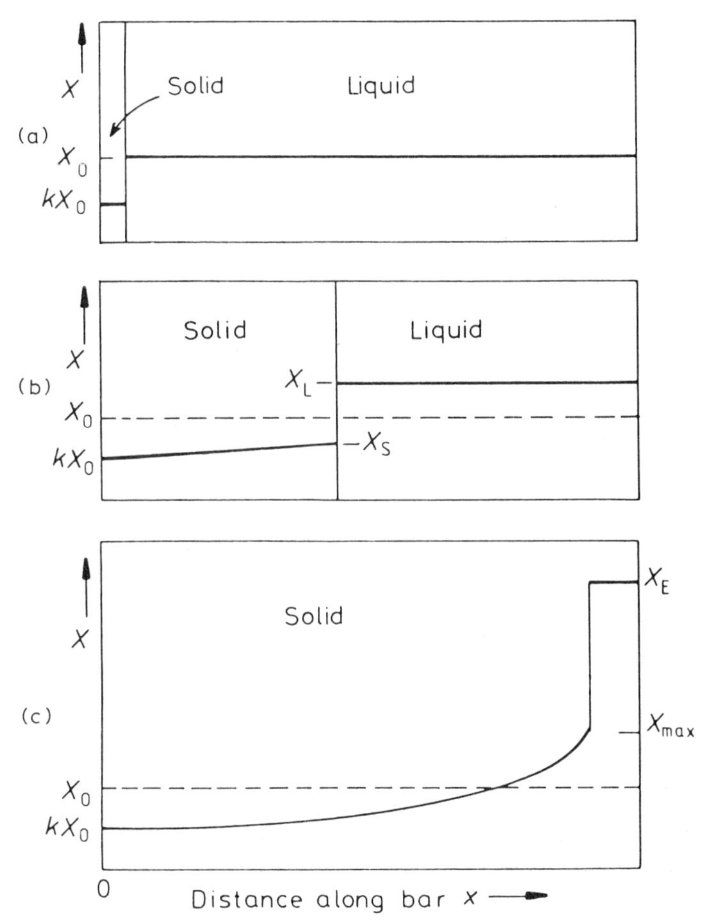

Figure 2.15 If there is perfect mixing in the liquid and no diffusion in the solid, the composition profile in a solidifying cylindrical bar (a) progresses as shown from (b) to (c). From Porter and Easterling[3]

the liquid becomes progressively richer in solute and solidification occurs at progressively lower temperatures. Now, since the assumption is that, initially, there is perfect mixing of the melt due to turbulence, 'local equilibrium' continues to exist at the solid–liquid interface, i.e. for a given interfacial temperature the compositions of the solid and liquid in contact with one another are given by the equilibrium phase diagram, *Figure 2.14*. However, since there is no diffusion in the solid the separate layers of solid retain their original compositions so that the mean composition of the solid is continuously *lower* than the solidus composition given by the phase diagram, i.e. the true solidus is defined by $\overline{X}_S$ in *Figure 2.14*. On the other hand, the liquid becomes progressively *richer* in solute and may even attain eutectic composition at the temperature T_E. The corresponding composition profiles for the volume element will thus progress as shown in *Figures 2.15* (*b*) and (*c*).

No diffusion in solid; diffusional mixing in liquid

If there is no stirring or convection in the melt, the solute rejected into the liquid during solidification can only be transported away from the solid–liquid interface by diffusion. There will thus be a build-up of solute ahead of the growing crystal, with a corresponding rapid increase in alloying content of the

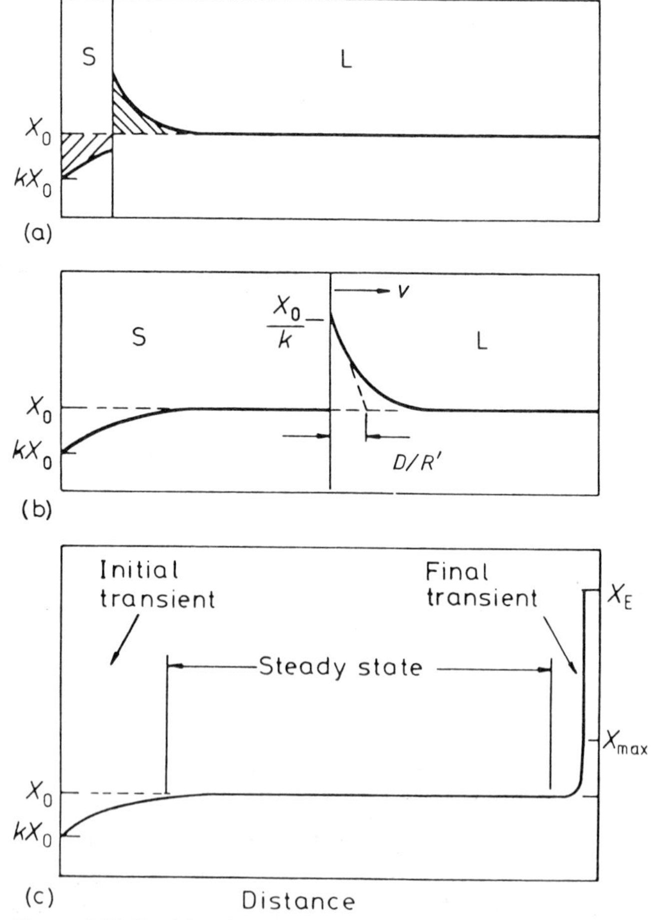

Figure 2.16 If mixing in the liquid is controlled by diffusion, the composition profile of a solidifying bar appears as the sequence (a) to (c). Provided the velocity of solidification remains uniform, the composition in the bar reaches a steady state until the final stage (c) when the last melt to solidify becomes more enriched. From Porter and Easterling[3]

solid that forms, *Figure 2.16*. During steady-state growth the concentration profile in the melt is such that the rate at which solute diffuses down the concentration gradient away from the interface is balanced by the rate at which solute is rejected from the solidifying liquid. *See* equation (2.12), where D is

$$-\frac{dX_L}{dx} D = R' (X_L - X_S) \tag{2.12}$$

the diffusivity of the melt, X_L and X_S are the solute concentrations of the liquid and solid in local equilibrium at the interface, dX_L/dx refers to the concentration gradient into the liquid and R' is the growth speed of the crystal (*see*, e.g., equation (2.3)). Note that dissipation of latent heat of solidification need not be considered in this equation because the rate of heat conduction is several orders of magnitude faster than solute diffusion and it is thus not important in alloy solidification.

If the diffusion equation is solved for steady-state solidification it can be shown that the concentration profile in the liquid ahead of the interface is given by equation (2.13). This equation indicates that X_L decreases exponentially

$$X_L = X_o \left\{ 1 - \frac{1-k}{k} \exp\left[-\frac{x}{D/R'} \right] \right\} \tag{2.13}$$

from X_o/k at $x=0$ (the interface) to X_o at some distance from the interface. The distance $\bar{x}$ roughly corresponds to the ratio of diffusivity to growth velocity, *see Figure 2.16* and equation (2.14). Savage[4] has estimated $\bar{x}$ for various welding

$$\bar{x} = \frac{D}{R'} \tag{2.14}$$

processes assuming an average solute diffusion coefficient of *ca.* 5×10^{-5} cm^2 sec^{-1}, and typical values of R' as given in *Table 2.1*. The values of $\bar{x}$ given are thus the predicted widths of the solute pile-up in front of the moving interface under steady-stage solidification. This suggests that, for example, in electro-

TABLE 2.1 Extent of diffusion controlled concentration gradient in welding processes (after Savage[4])

Processes	*R' (cm sec^{-1})*	*$\bar{x}$ (cm)*
Electroslag	*ca.* 10^{-3}	*ca.* 0.25
Manual metal arc	*ca.* 0.2	*ca.* 10^{-3}
Electron beam	*ca.* 5	*ca.* 5×10^{-5}

slag welding the solute enrichment zone at the weld centre where the growing interfaces meet can be up to *ca.* 5mm in width. Thus, in the case of solutes of low k, e.g. sulphur in steel, this can obviously give rise to hot cracking problems*. Even normal arc welding is seen from *Table 2.1* to exhibit solute-enriched regions at the weld centre line of *ca.* 10^{-2} mm, which is still a few orders of magnitude more than the width of a grain boundary.

* *See Table 4.1*

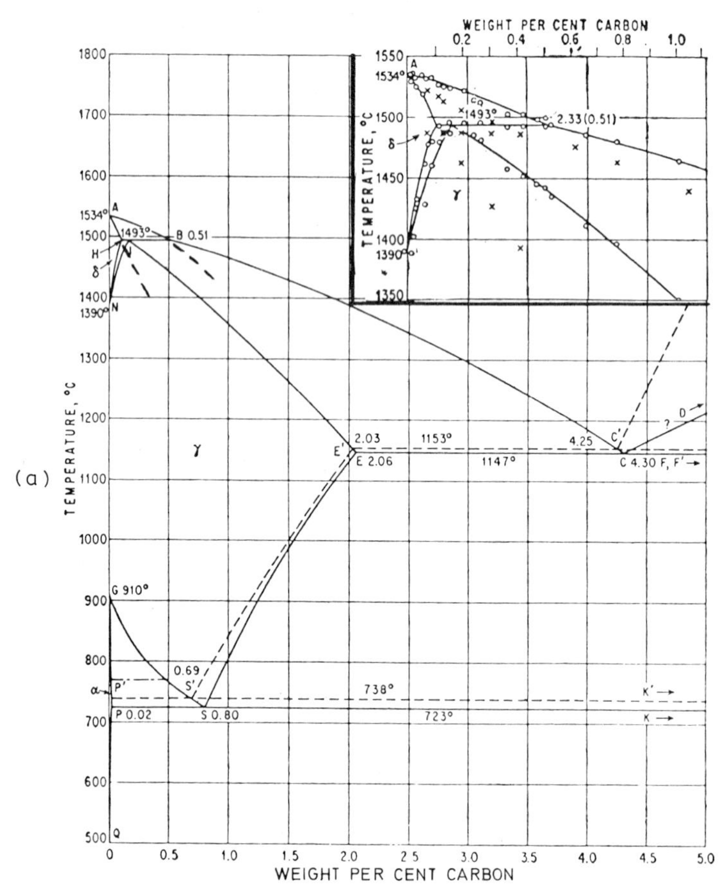

Figure 2.17 (a) Comparison between the solidification ranges of dilute alloys of Fe–C. After Hansen, M., *Constitution of Binary Alloys*, McGraw-Hill, 1958

As indicated by *Figure 2.16*, in the final stages of solidification, the 'bow-wave' of solute is compressed into a comparatively small volume of liquid and the solute composition is thus expected to rise rapidly. In terms of weld solidification, this suggests that the last stages of weld solidification are associated with the highest level of segregation and that the segregation effect is likely to be greatest for those patterns of converging crystal growth that are typical of high-speed welding (*see Figure 2.17* (*b*)).

Factors that can affect the degree of segregation are worth discussing in relation to *Figure 2.14*. A simple binary system is considered above, but most weld metals contain a number of impurities and alloying elements. Indeed, because of dilution alone the composition of the weld metal typically lies between those of the base metal and the filler wire. As a general rule, alloying elements or impurities that tend to *widen* the solidification range, i.e. $T_1 - T_3$, or that possess a low partition coefficient, k, will increase the amount of segregation. This is illustrated in *Figure 2.17* in which the solidification ranges for (*a*) Fe – C and (*b*) Fe–S are compared. The more extensive solidification

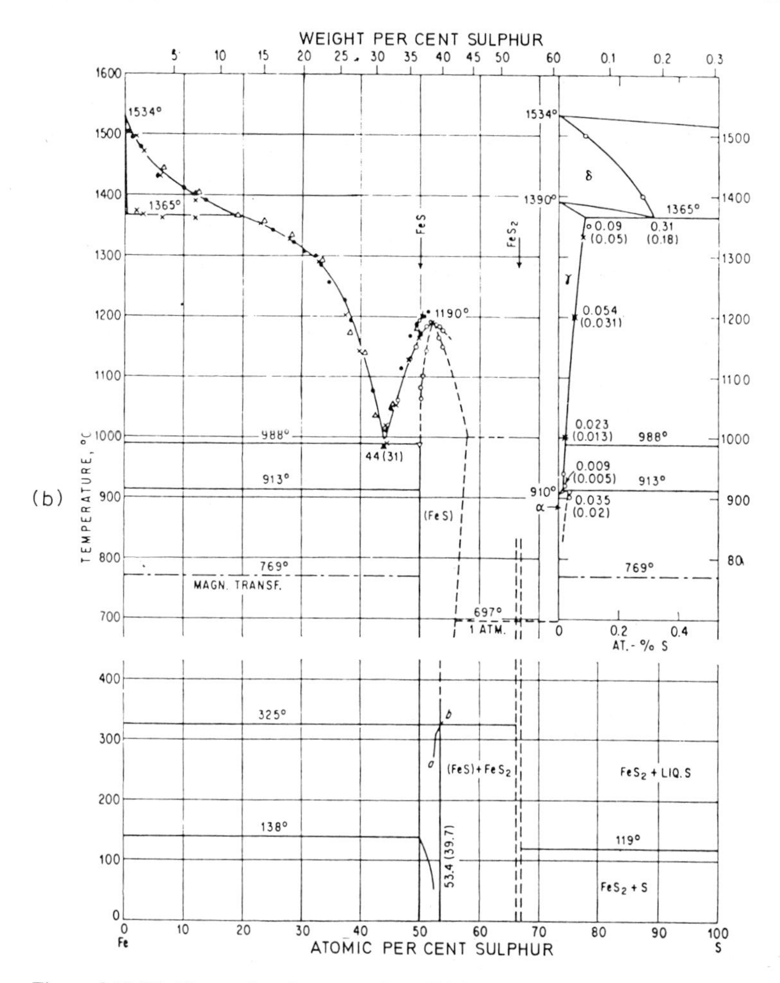

Figure 2.17 (b) Comparison between the solidification ranges of dilute alloys of Fe–S. Note that the two-phase liquid + solid field extends to much higher compositions than in (a). After Hansen, M., *Constitution of Binary Alloys*, McGraw-Hill, 1958

range of sulphur in iron (about 12 wt% S compared with 0.5 wt% C) make this element particularly dangerous in connection with hot-cracking of welds. The effect of rapid cooling, as in welding, may even depress the final solidification temperature, T_E. *Figure 2.17 (a)* illustrates this for the case of steel, where it is seen that if, due to rapid cooling, solidification has not had time to go to completion at the peritectic temperature, a *quasi-equilibrium* situation is realized in which the δ-liquidus and solidus lines are effectively extended. This results in an increase in the amount of segregation since the liquidus extends its composition range.

It was assumed at the beginning of this section that no diffusion occurs in the solid. This is not a valid assumption, particularly in the case of interstitial alloys, such as steel, in which changes in carbon content may occur in the base metal close to the transition line[5]. For a more detailed treatment of this problem, *see,* e.g., ref. 3.

Cellular and dendritic solidfication in welds

It was suggested in the previous section that solidification of welds progresses from a condition characterized by *turbulence* — hence good mixing of alloying elements — in the melt, to one in which mixing is more controlled by *diffusion* of alloying elements in the melt. In terms of *Figure 2.16,* this results in a change in concentration profile ahead of the solid–liquid interface in which the value of D/R' progressively decreases. Another variable that should be considered when trying to predict the types of microstructure developed in weld solidification is the change in temperature gradient across the melt, since this also reduces as D/R' decreases.

Consider steady-state solidification at a planar interface, *Figure 2.18,* in which both the concentration profile and the temperature gradient are shown as a function of distance from the interface, x. Now, referring first to the equilibirium diagram *Figure 2.14,* if the solute concentration of the alloy

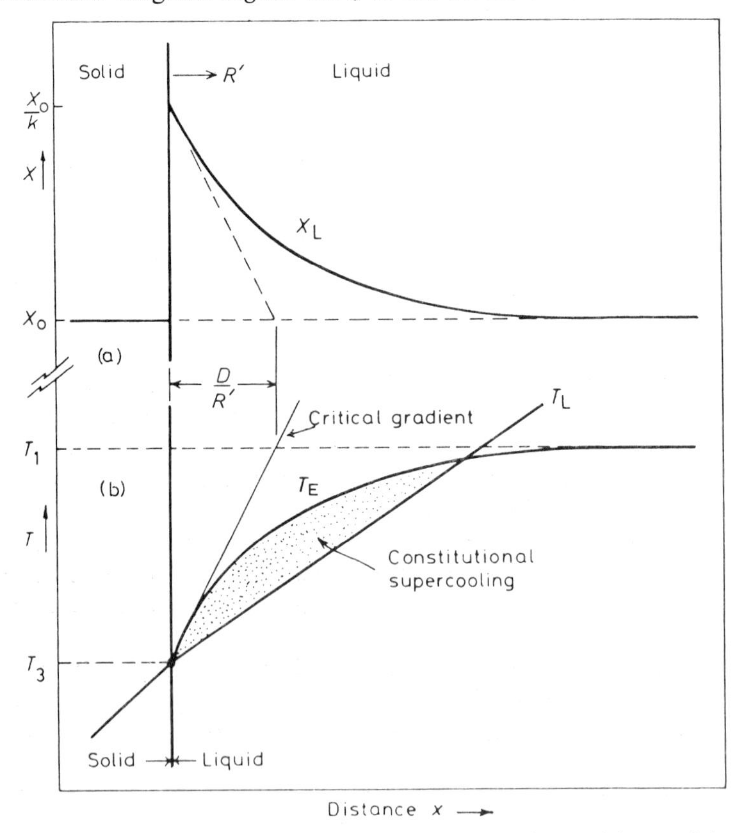

Figure 2.18 Constitutional supercooling occurs when solute enrichment of the liquid just ahead of the growing solidification front causes the solidification temperature of the liquid to be lowered. The figure shows that constitutional supercooling is thus dependent on the temperature gradient in the melt, T_L. From Porter and Easterling[3]

increases its solidification temperature, T_E, decreases, at least up to the eutectic composition. Thus, the temperature gradient, T_E, in *Figure 2.18* refers to the solidification temperatures corresponding to the enriched liquid ahead of the interface, X_L. However, the actual temperature gradient in the melt is determined by the position of the heat source relative to the interface, and as stated above, this varies as solidification to the weld centre line progresses. Assume a temperature gradient, T_L, at some moment in time, as given in *Figure 2.18,* where $T_L = dT/dx$ in the liquid. If the temperature of the melt lies *below* the solidification temperature of the enriched liquid, this part of the melt must be *supercooled.* In other words, this liquid (shaded region in *Figure 2.18*) remains liquid in spite of the fact that its temperature is below the equilibrium solidification temperature. Since supercooling arises from compositional or *constitutional* effects, this phenomenon is known as *constitutional supercooling.* How constitutional supercooling affects solidification microstructure is now discussed.

Consider an initially planar interface, growing under conditions of constitutional supercooling (*Figure 2.19*), in which a small nodule has moved ahead

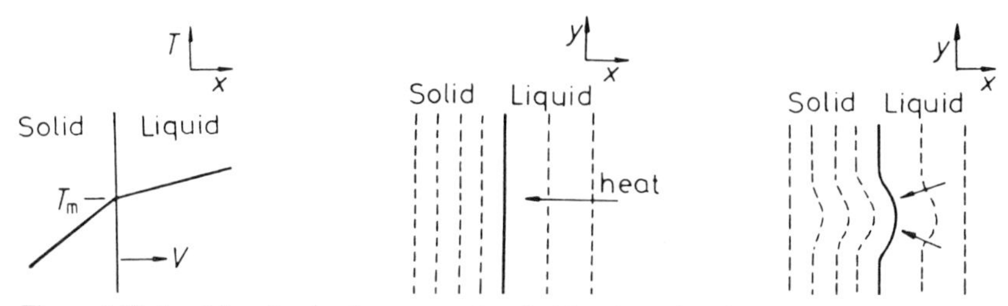

Figure 2.19 A solid protrusion that grows ahead of the planar interface normally melts back as it enters hotter liquid. However, if the liquid ahead of the interface is constitutionally supercooled, the protrusion tends to stabilize instead

of the interface. Despite the fact that the tip of the nodule is hotter than the planar interface it may still be stable, since its temperature remains below the liquidus temperature. The reason for this can be understood by reference to *Figure 2.18,* in that because of solute enrichment the liquid just ahead of the interface has a *higher* solidification temperature. Thus, only if the actual temperature in the liquid is higher than T_E will the nodule melt again. It is useful to describe the critical temperature gradient for constitutional supercooling in terms of the growth speed of the interface and the welding conditions.

Under steady-state growth, the critical gradient (*see Figure 2.18*) is given by the ratio of the equilibrium solidification range of the alloy (T_1–T_3) and D/R', so that a stable planar interface can develop when equation (2.15) holds.

$$T_L > \frac{(T_1 - T_3)}{(D/R')} \tag{2.15}$$

Rearranging the experimental variables, the condition for constitutional supercooling to occur is given by equation (2.16) or in terms of the welding speed and

$$\frac{T_L}{R'} < \frac{T_1 - T_3}{D} \tag{2.16}$$

position of the interface relative to the heat source, equation (2.17) is obtained

$$\frac{T_L}{v \cos \Theta} < \frac{T_1 - T_3}{D} \tag{2.17}$$

from equation (2.2). On the basis of this relationship, it can be stated that planar front solidification is difficult to maintain in alloys with a large solidification range, when welding speeds are high, or when the growth front tends to follow behind the heat source ($\Theta \rightarrow 0^o$, *see Figure 2.6)*.

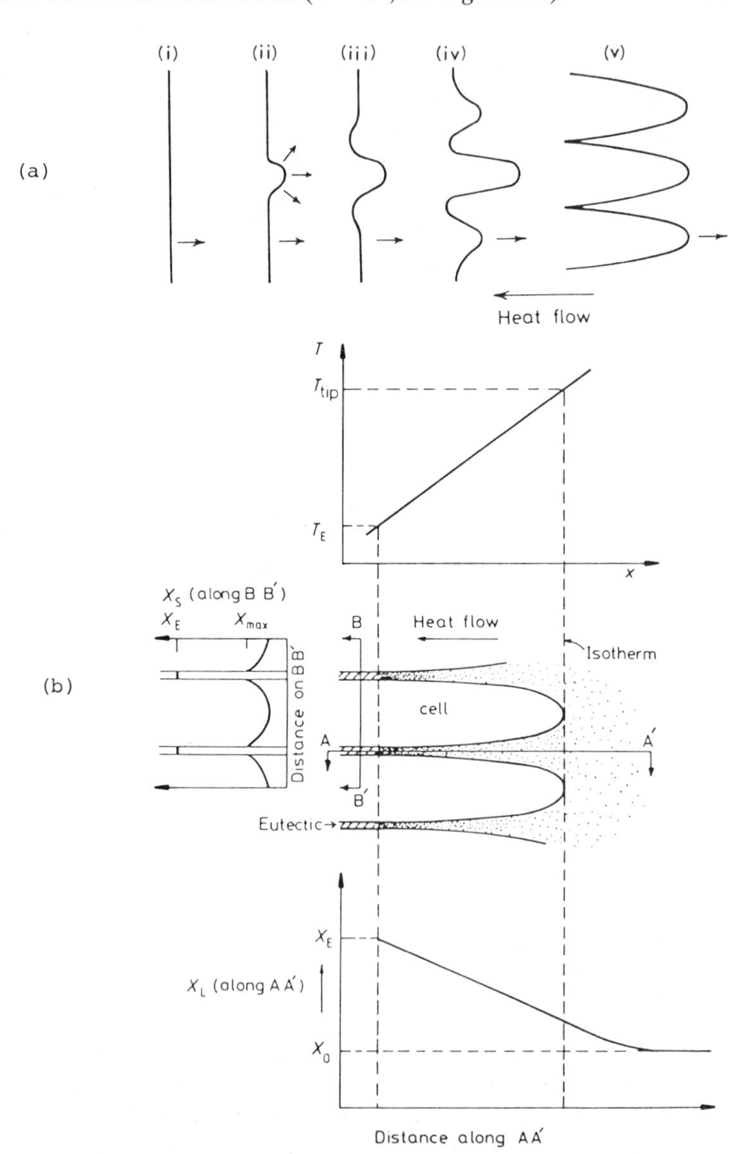

Figure 2.20 (a) The breakdown of a planar interface into cells. (b) Temperature and solute distributions associated with cellular solidification. From Porter and Easterling[3]

Thus, in weld solidification, where T_L is initially high and Θ is low, epitaxial growth from the base metal is likely to occur initially by the development of a planar growth front. However, as T_L decreases and Θ decreases, constitutional supercooling is expected to occur, resulting in a transition from planar to cellular growth. This transition is illustrated in *Figure 2.20* (*a*) and is worth studying in more detail.

The formation of the first protrusion *Figure 2.20* (ii), causes solute to be rejected laterally and to collect at its root. This effectively lowers the local solidification temperature at the root which causes recesses to form, *Figure 2.20* (iii), which in turn trigger the formation of other protusions, *Figure 2.20* (iv). Eventually the protrusions develop into long arms of *cells* that grow approximately parallel to the direction of maximum heat flow, *Figure 2.20* (v). The solute rejected from the solidifying liquid thus concentrates between the cell walls which then solidify at lower temperatures. Meanwhile, the tips of the cells, which continue to grow into the warmer melt, are less enriched with solute. The resulting intercellular solute enrichment is shown in *Figure 2.20* (*b*) as a function of solidification temperature and distance ahead of the interface.

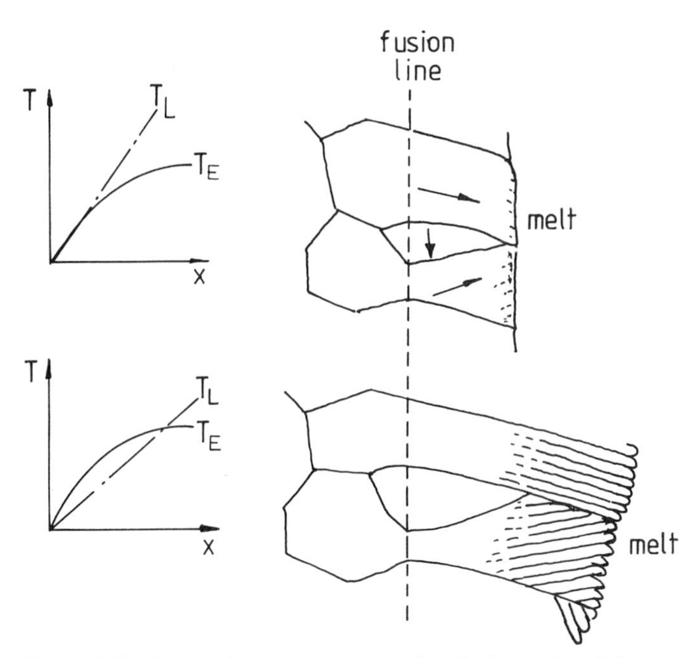

Figure 2.21 Changes in temperature gradient in the melt and the resulting onset of constitutional supercooling bring about the planar→ cellular transition. The arrows indicate the crystal growth directions.

As indicated in *Figure 2.20* (*b*), the enrichment between cells may attain the eutectic composition of the alloy. The interaction between temperature gradient, cell shape and solute segregation is shown in *Figure 2.21*. The planar-to-cellular transition described is, in fact, fairly typical of many welds and is illustrated in *Figure 2.22* (*see also Figure 2.8*). It is found that in most metals the cellular spacing, *s*, is related to the crystal growth rate, R', and the temperature gradient, T_L, by equation (2.18) or, in terms of the welding speed v, by equation (2.19).

$$s = \frac{1}{R' T_L} \tag{2.18}$$

$$s = \frac{1}{(v \cos \Theta)\, T_L} \tag{2.19}$$

Thus, for a given welding process, factors which influence T_L affect the cell spacing. For example, a high melting point alloy, such as steel, which has a relatively poor thermal conductivity typically has a finer cell spacing, than, say, aluminium with its lower melting point and better conductivity. It is also apparent that higher welding speeds give finer spacings.

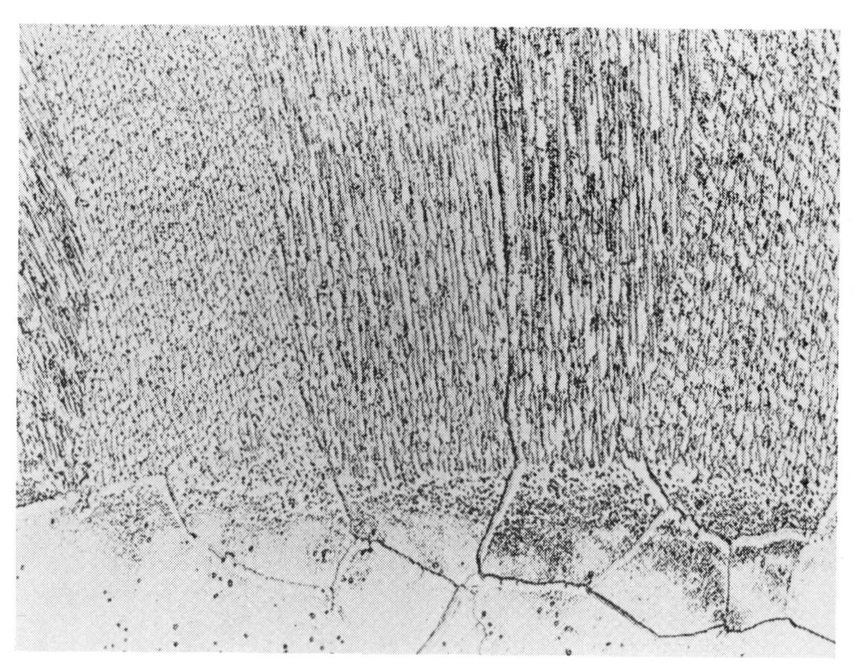

Figure 2.22 The planar → cellular transition in a TIG-welded Al – 4.5 Cu alloy. After Davies and Garland[1] (×260; then reduced by one-fifth in printing)

Cell spacing also affects the degree of segregation at the cell or dendritic boundaries, finer spacings giving less segregation. In determining the degree of segregation, in fact, the important factors are:

1. Density and spacing of cell boundaries.
2. The segregation or partition coefficient, k, of the solute.
3. The total amount of solute present.

The latter criterion is of particular importance in high alloyed deposits such as stainless steel.

Cellular microstructures are only stable for a certain range of T_L or Θ and as the temperature gradient, or Θ, reduces so the walls of the *primary cells* become unstable and develop *secondary arms,* and at a later stage *tertiary arms,* i.e. a *dendritic* structure, are developed, *Figure 2.23.* Although dendrite formation is not fully understood, it is thought that the same phenomenon of constitutional supercooling that governs the growth of primary protrusions from a planar interface also governs the formation of secondary and tertiary arms. However, in this case, there is no temperature gradient perpendicular to the primary arms. Thus, the arm spacing developed is probably that which reduces the constitutional supercooling in the intervening liquid to its lowest

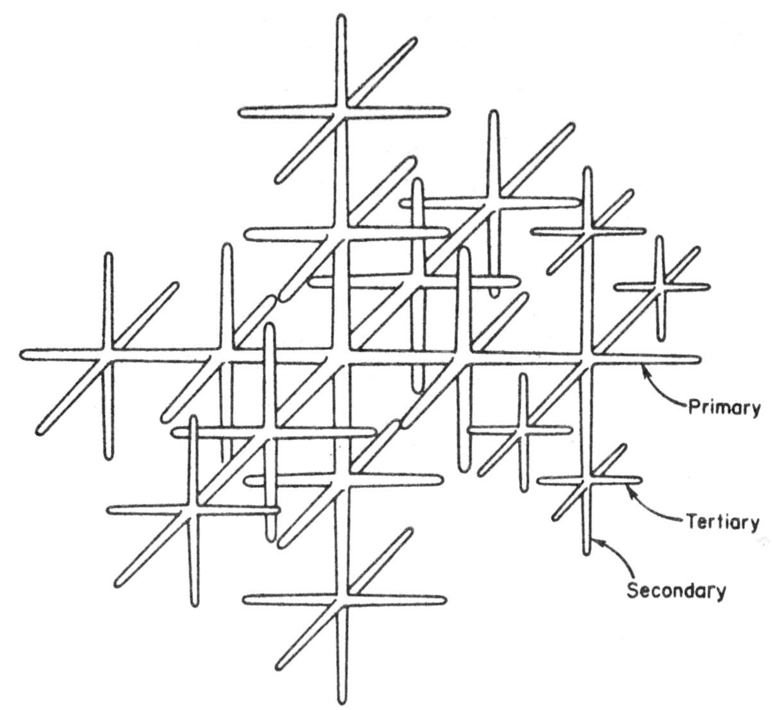

Figure 2.23 A fully developed dendrite. The primary, secondary and tertiary arms all correspond to <100> growth directions in cubic metals

level. This is consistent with the experimental observation that both cell and dendrite arm spacings decrease with increasing cooling rate, since higher cooling rates allow less time for lateral diffusion of the rejected solute and thus require smaller arm spacings to avoid constitutional supercooling. This implies that in weld solidification, which is a high cooling rate process, the cell spacing is relatively fine. Indeed, cell structures in weld deposits are typically finer than in castings.

The transition from cellular to dendritic structures is observed in weld solidification, particularly in high speed welding where dendrites form at the centre line of the weld, where Θ and T_L are at their lowest. An example of dendrites in TIG-welded Monel is shown in *Figure 2.24.* Experimentally, it is found that equiaxed dendritic regions develop at the highest speeds and in the

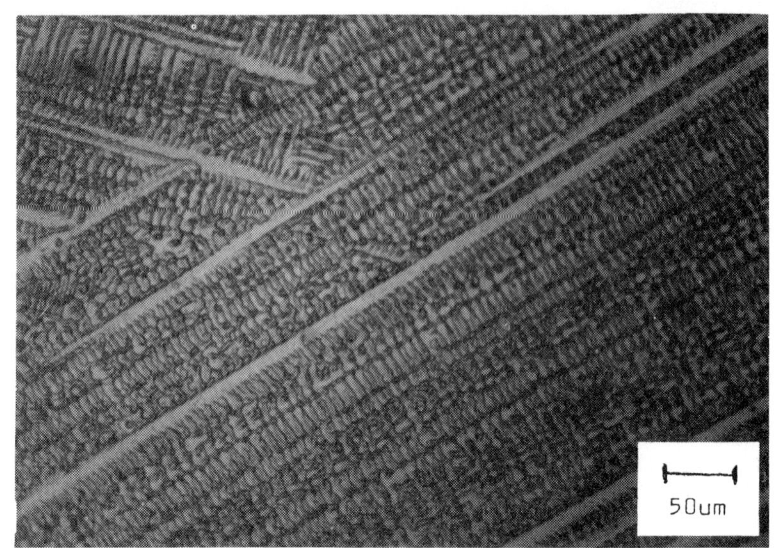

Figure 2.24 Dendritic growth in TIG-welded Monel. Courtesy of Gudrun Keikkala, University of Luleå

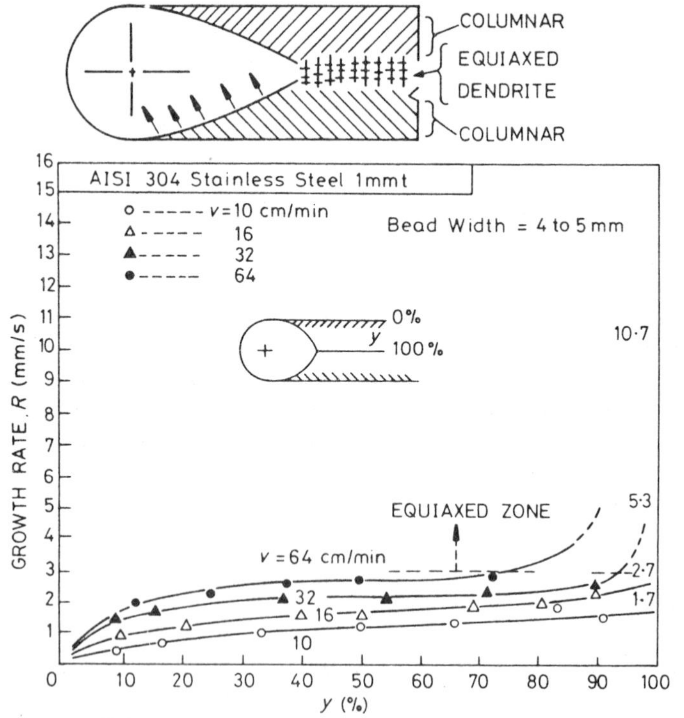

Figure 2.25 These measurements of T.Senda, (*TIG-welded Stainless Steel*, Technical Report, Osaka University, **20,** 932, 1970) show that the last stage of solidification is associated with high growth speeds and low temperature gradients, giving rise to dendritic structures at the weld centre line. *y* refers to the percentage of solidification. Note that the highest welding speeds correspond to the highest crystal growth rates and earlier onset of dendrite formation

final stages of solidification or in high alloy deposits, all as expected on the basis of the above arguments. This is illustrated in *Figure 2.25*.

If the heat source is removed at the end of the weld run, final solidification of the weld is in the form of an elliptically shaped pool which produces a 'crater effect'. Since solidification of this crater is associated with rapid crystal growth, large segregation and a low temperature gradient, constitutional supercooling is likely to be high and the final structure is thus dendritic.

The various contributions of temperature gradient, alloying, crystal growth rate, etc., on the solidification structures developed in welds are illustrated schematically in *Figure 2.26.* It is concluded that application of the

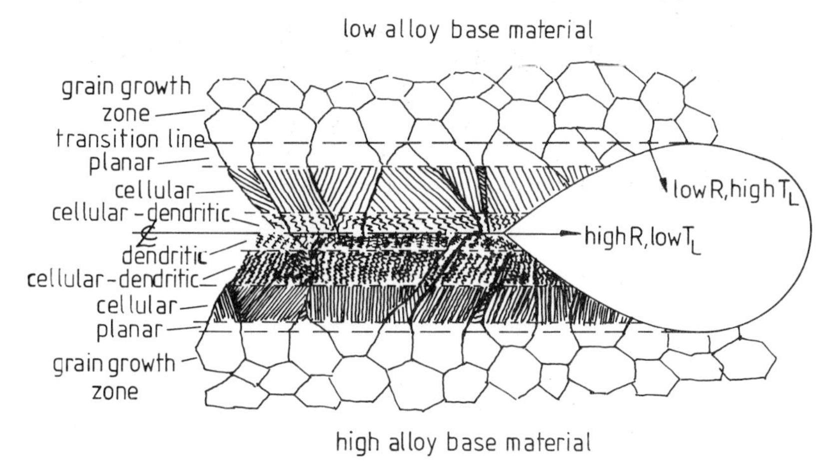

Figure 2.26 Schematic illustration of the various types of growth products developed during the solidification of weld metal as a function of alloy composition, crystal growth rate and temperature gradient in the melt

principles of solidification theory to the special conditions of fusion welding gives a reasonable understanding of the effect of the various welding and material parameters on the type of solidification structure developed as well as of the factors that are important to segregation. An important conclusion, too, is that the cellular–dendrite structure in welds is relatively fine compared with most castings. This is important in terms of toughness since, unlike castings, welds often form an integral part of a strong, tough structural component.

Refining weld structures

Research is in progress to develop techniques for refining welds by, e.g., magnetic stirring, ultrasonic cavitation or the use of chemical inoculators. For a useful review *see* ref. 1. This approach is of particular importance in high energy input welds such as submerged arc and electroslag, where the grain structure is relatively coarse. It is claimed that grain refinement not only results in a homogeneous equiaxed grain structure, but even reduces centre line segregation. Currently, however, there are still very few industrial applications of grain refinement in weld metals, except possibly in the Soviet Union where some progress has been reported with electromagnetic stirring of electroslag welds[6]. Limited success has also been reported using this technique on GTA welds of aluminium alloys[7]. Perhaps the simplest technique involves the use of inoculators, but there are problems due to the negative effect of these additional 'impurities' on weld toughness (*see* e.g., ref. 8).

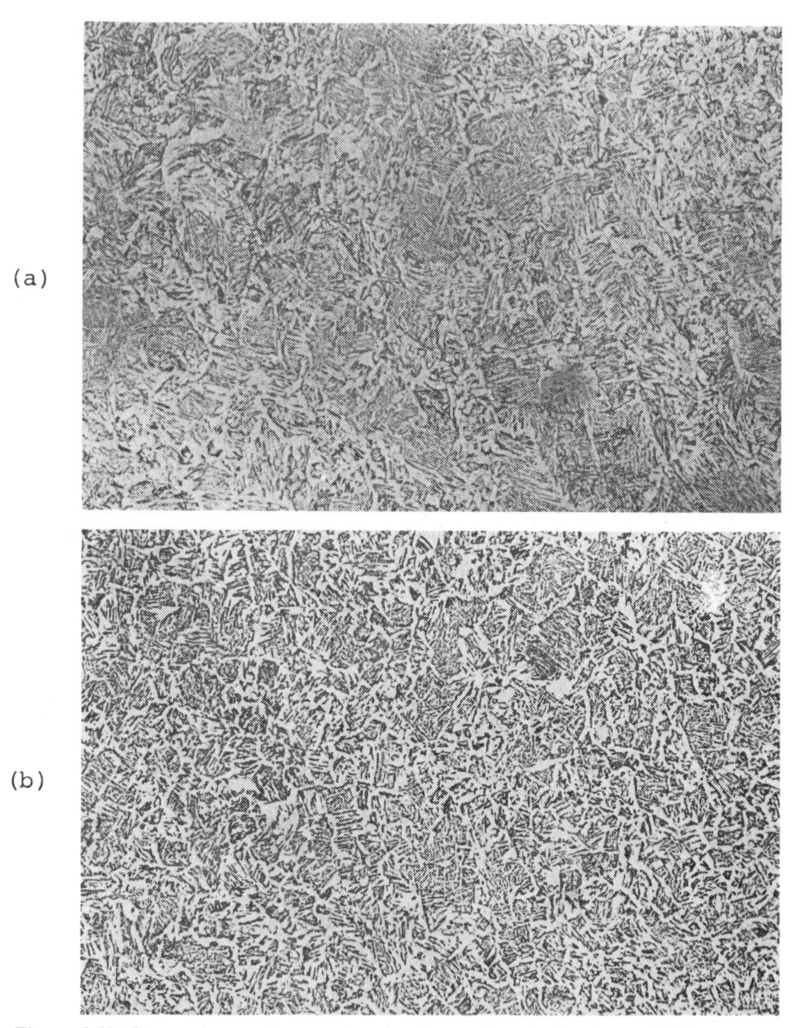

Figure 2.27 Comparison between microstructures of an electroslag weld (a) without and (b) with ultrasonic vibration (20 KHz) solidification. Both micrographs are from the weld centre lines. Courtesy of Hans Liljenfeldt, University of Luleå (× 40; then reduced by one-fifth in printing)

Limited progress has been achieved with the use of ultrasonic vibration in electroslag welding and an example of the refinement of structure achieved using this approach is shown in *Figure 2.27*. The main problem in this case is to find a suitable material which can withstand the effects of cavitation at the horn tip. Currently, experiments are in progress in which ceramic coatings at the tip are being tried; the body of the horn is cooled by the water-cooled copper jacket of the electroslag process[9].

It is evident from *Figure 2.27* that the effect of cavitation is to reduce the grain size quite considerably. The mechanism by which this refinement occurs is not properly understood. Evidently, the large pressure changes (expansion and contraction), that result from cavitation somehow cause new crystals to be nucleated at the growing solidification interface, thereby avoiding the tendency for epitaxial solidification. Whether these new crystals form as a result of the *contraction* cycle by the nucleation of solid embryos or because of the mechanical disturbances on the growing dendrites during the *expansion* cycle remains, however, to be proven.

Phase transformations during cooling of the weld metal

During cooling of the weld metal from its solidification temperature to the ambient temperature, additional modifications in microstructure and solute distribution are likely to occur even in single phase metals. In polyphase metals, however, phase transformations or modifications in phase distribution do occur. The most important of this group of metals are the transformable (carbon and low alloy) and stainless steels and these transformations are dealt with below.

Kinetics of phase transformations

The rapid cooling rates of weld metals determine that transformations which involve the solid phases do not usually occur under equilibrium conditions. There is, in fact, a nucleation problem in that a certain minimum energy is needed in order to accommodate the surface and strain energies of the new phase. This is illustrated in *Figure 2.28,* in terms of the molar free energy of the transformation, ΔG, equation (2.20). ΔG_V is the chemical free energy of the

$$\Delta G \geqslant - \Delta G_V + \Delta G_S + \Delta G_E - \Delta G_D \qquad (2.20)$$

nucleus and is negative because it assists the transformation. ΔG_S refers to the increase in surface energy between γ- and α-phases, and ΔG_E is the increase in strain energy due to the lattice dilatation involved. T_o in *Figure 2.28* refers to

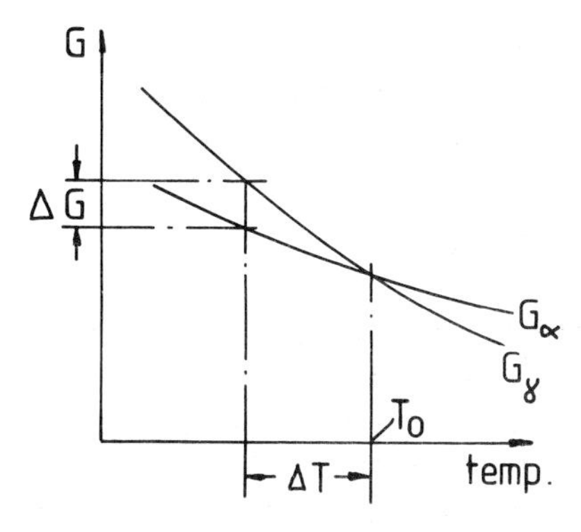

Figure 2.28 The molar free energies of the two solid phases γ and α as a function of temperature. Owing to problems of nucleation, a certain minimum free energy, ΔG and an undercooling, ΔT, are needed to initiate the $\gamma \rightarrow \alpha$ transformation

the temperature at which the free energy of the α- and γ-phases are equal. ΔG_D implies that the transformation is *heterogeneous* and refers to the energy donated to nucleation by an existing heterogeneity in the high temperature phase. ΔG_D can thus refer, for example, to a grain boundary. ΔT in *Figure 2.28* refers to the undercooling of the alloy below T_o necessary for ΔG to be sufficiently large for the new phase to nucleate. Since some diffusion or redistribution of solute atoms is needed before the nucleus has the size and composition appropriate to the new phase, and because this redistribution of solute atoms requires a certain time, it is apparent that ΔT is dependent on the *cooling rate* of the alloy. It is important to see how ΔT is related to the nucleus size and to ΔG.

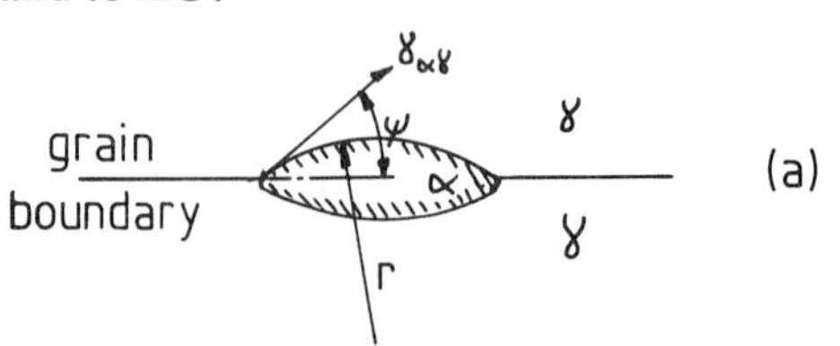

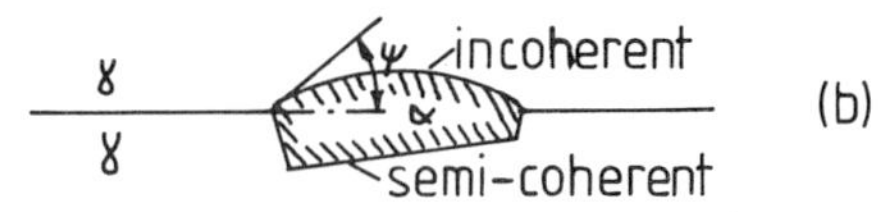

Figure 2.29 The shape of a second phase growing on a grain boundary depends on the relative interphase energies as shown for (a) an incoherent particle and (b) a semi-coherent particle

Consider a new phase, α, nucleating on the grain boundary of γ, *Figure 2.29 (a)*. Ignoring the strain energy term, which is rather small in a diffusional transformation, the nucleus tries to adjust its shape to minimize surface energy. The optimum shape of a nucleus in which the γ/α phase boundary is *incoherent* (i.e. a high angle boundary), is described by equation (2.21), where γ refers to

$$\cos \psi = \frac{\gamma_{\gamma/\gamma}}{2\gamma_{\alpha/\gamma}} \tag{2.21}$$

the surface energy between phases α and γ. This equation may, in fact, be derived from equation (2.5). It may be shown that the critical radius, r^*, of the embryo is given by equation (2.22), where r^* is defined in *Figure 2.13*.

$$r^* = \frac{2\gamma_{\alpha/\gamma}}{\Delta G_V} \tag{2.22}$$

If the interface between α and γ is oriented so that it is coherent or semi-coherent (*Figure 2.29 (b)*), i.e. such that a low energy *orientation relationship* exists between the phases, then since $\gamma_{\alpha/\gamma}$ (coherent) $< \gamma_{\alpha/\gamma}$ (incoherent), r^* could be correspondingly smaller. Both types of interface occur in eutectoidal transformations, often as shown in *Figure 2.29 (b)*. Thus, the ability of the grain boundary (or any other defect) to reduce ΔG depends on its potency as a nucleation site, i.e. as the ratio $\gamma_{\gamma/\gamma}/\gamma_{\alpha/\gamma}$. The most potent nucleation sites in decreasing order are:

1. Free surfaces.
2. Grain corners.
3. Grain boundaries.
4. Inclusions.
5. Dislocations, including stacking faults.
6. Vacancy clusters.

Some of these sites are shown in *Figure 2.30*.

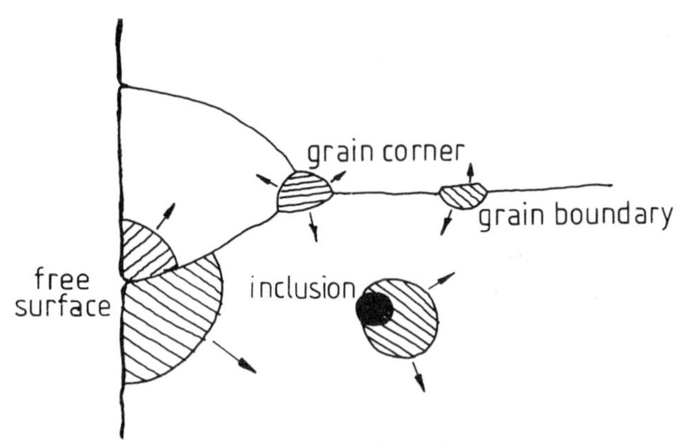

Figure 2.30 The most potent sites for the nucleation of a new phase. The importance of a given site in terms of total volume fraction depends also on the number of sites available

In practice the relative amount of transformation as a whole is also a function of the total number of sites available. If the concentration of nucleation sites is C_1 per unit volume of metal then the nucleation rate, N, is given by equation (2.23), where $\omega \exp(\Delta G_m/kT)$ refers to the activation energy

$$N = \omega C_1 \exp\left(\frac{\Delta G_m}{kT}\right) \exp\left(\frac{\Delta G^*}{kT}\right) \text{ nuclei m}^{-3}\text{ s}^{-1} \qquad (2.23)$$

for atomic migration to form the nucleus, ω being the vibrational frequency of atoms and k the Boltzmann constant. Equation (2.23) can be plotted as a function of temperature, as shown in *Figure 2.31*. The nucleation rate is strongly dependent upon temperature, as shown by equation (2.23), which gives rise to the C-shape behaviour of phase nucleation and growth. This relationship can be plotted, instead, as a function of temperature and time, as shown in *Figure 2.32* (*a*). In this case boundaries are plotted which correspond to 1% and 99% transformations, these being more convenient to measure experimentally than 0 and 100%. Now, if C^* is the concentration of critical sized nuclei, then equation (2.24) can be used, where C_o is the number of atoms

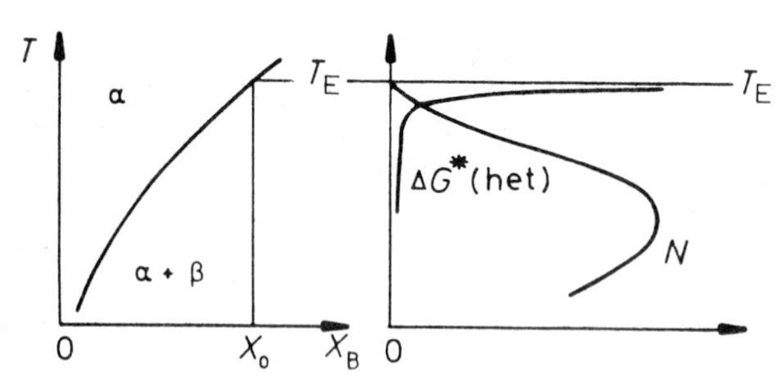

Figure 2.31 The rate of heterogeneous nucleation during precipitation of a new phase, α, in an alloy X_o, as a function of undercooling

$$C^* = C_o \exp\left(\frac{\Delta G^*}{kT}\right) \tag{2.24}$$

per unit volume of the new phase. If each nucleus can be made supercritical at a rate of f per second, equation (2.23) can be rewritten as equation (2.25), where $f = \omega \exp(\Delta G_m/kT)$ and is thus temperature dependent. This is illustrated in *Figure 2.32* (*b*) by plotting f for two different temperatures, T_1 and T_2.

$$N = f\,C^* \tag{2.25}$$

If the cooling rate is increased, resulting in larger values of ΔT, other types of transformation product can develop based on different morphologies, modes of growth, or compositions and lattice structures. Each new product has

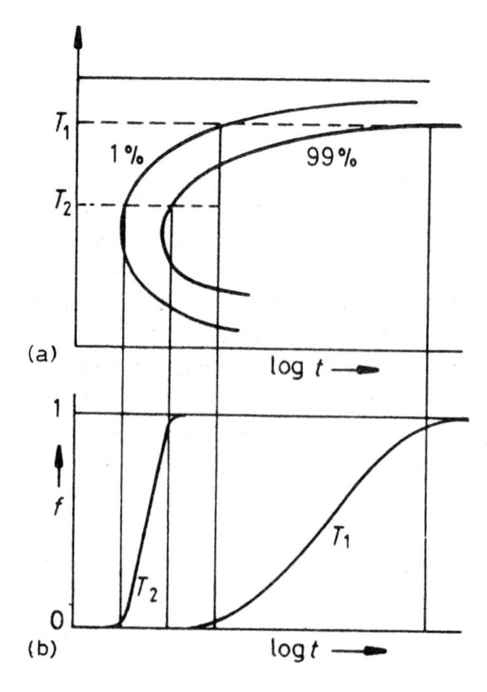

Figure 2.32 The percentage of transformation of a new phase as a function of transformation temperature

its own characteristic C-curve, as in *Figure 2.32* (*a*); the maxima are merely displaced to lower temperatures. When several types of transformation product are concerned, as with steel, the diagram may appear as in *Figure 2.33*. This is known as a *continuous–cooling–time* (CCT) diagram, to denote transformations that occur in samples being continuously cooled, rather than developing under isothermal conditions in which case a *time–temperature–transformation* (TTT) diagram is used. The use of CCT diagrams is clearly more appropriate to the weld situation. It is seen from *Figure 2.33* that a

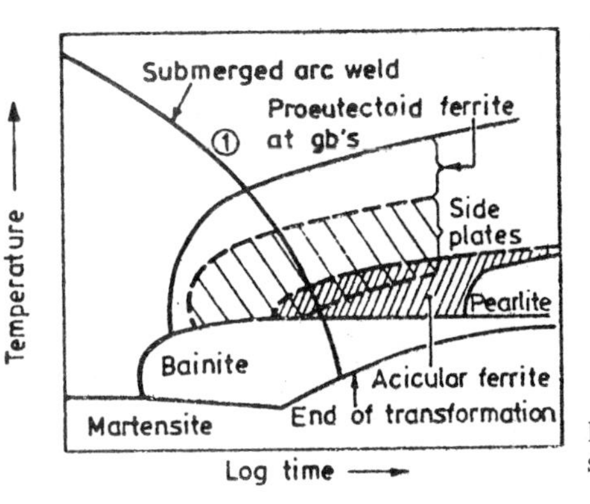

Figure 2.33 Schematic CCT diagram for a steel weld metal. After Abson and Dolby[10]

number of possible transformation products may appear in a sample at ambient temperature depending upon the cooling rate. Thus, for the cooling curve shown, which corresponds to the cooling of the weld metal, the final structure would consist of a mixture of:

1. Proeutectoid ferrite.
2. Widmanstätten side plates.
3. Acicular ferrite.
4. Bainite.

in order of increasing undercooling below the A_3 temperature.

The factors that determine the final volume fraction of the various transformation products are the nucleation rate, the growth rate, the density and dispersion of nucleation sites, the overlap of diffusion fields from adjacent products and the impingement of adjacent volumes. Some of these factors are illustrated in *Figure 2.34* which shows ferrite growth and impingement.

If the cooling of the metal is interrupted and held at such a temperature that the growth of the new phase is allowed to go to completion, the volume fraction, V_f, increases exponentially with time. Thus, 50% V_f is given by equation (2.26) and the time needed to obtain this amount is given by equation

$$V_{f(0.5)} = \exp(-0.7) \tag{2.26}$$

(2.27), where $\dot{G}$ is a function of the nucleation rate, $\dot{N}$, and the rate of growth,

$$t_{0.5} = \frac{0.7}{\dot{G}^{1/n}} \tag{2.27}$$

v, of the new phase; n is a numerical exponent with values between 1 and 4, depending on the *nucleation mechanism,* and it is independent of temperature. This equation shows that if $\dot{G}$ is large, the new phase grows very rapidly. $\dot{G}$ is highly sensitive to temperature because of its relation to $\dot{N}$ and even v is a temperature-dependent factor. For the case shown in *Figure 2.34* it can be shown[3] that equation (2.28) is applicable. The factors that affect $\dot{N}$ are the

$$t_{0.5} = \frac{0.9}{\dot{N}^{1/4}\ v^{3/4}} \tag{2.28}$$

number and type of nucleation sites and undercooling. Those that affect v are the type of interface (degree of coherency) and amount of alloying if changes in composition are required by the new phase. Thus, an important factor is the grain size of the parent phase because of its effect on $\dot{N}$. Another nucleation site

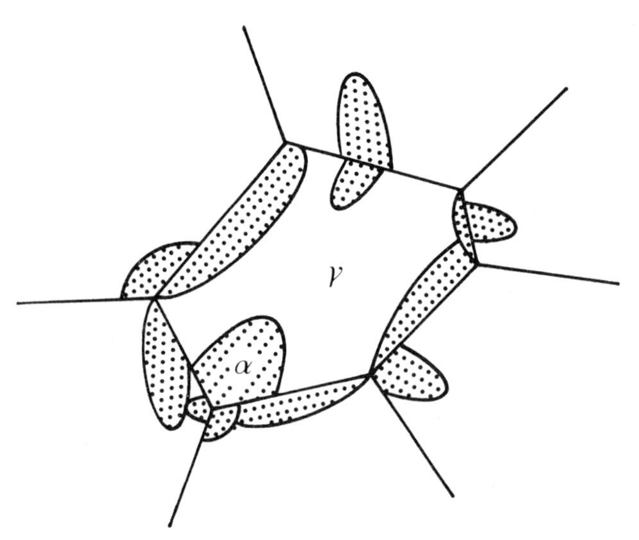

Figure 2.34 The nucleation, growth and impingement of ferrite at grain corners and boundaries

of interest in weld metals is the slag inclusion, and in this case the size, type and number of inclusions are all of relevance if r^* (equation (2.22)) is to be achieved and if the volume fraction arising from this source is to be meaningful. In fact, recent work suggests that in weld metals with oxygen levels in the high range (*ca.* 0.03—0.06%), ferrite nucleation at slag inclusions is meaningful[10], *see* p. 96 for a treatment of this.

So far only transformations in which nucleation and growth are controlled by diffusional processes, i.e. the long range interchange of atomic position by interstitial or vacancy – solute pair diffusion, have been discussed. However, as seen in the schematic CCT diagram (*Figure 2.33*), if the cooling rate is sufficiently rapid, *martensite* can form instead.

Martensite nucleates and grows by a *diffusionless shear mechanism* in which no single atom moves by more than one lattice spacing. Therefore, martensite has the same composition as the parent austenite. Martensite formation is illustrated in *Figure 2.35,* which shows that the new phase has formed by the shearing of austenite from one lattice configuration to another, while maintaining a high degree of coherency at the γ/α' interface. Thus, in this case the elastic term, ΔG_E, in equation (2.20) is much larger than the surface energy term, ΔG_S, since $\gamma_{\alpha'/\gamma}$ is small. This means that 'surfaces', such as grain

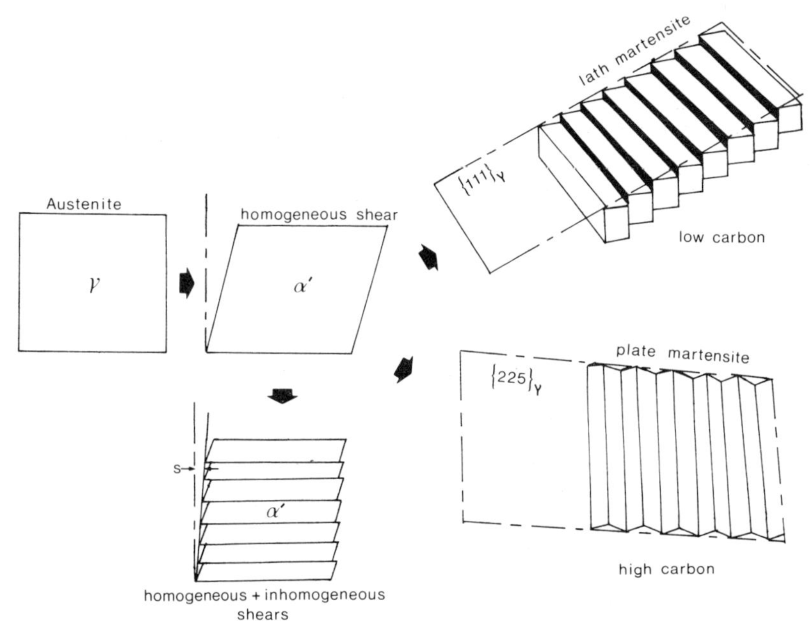

Figure 2.35 The morphologies of martensite. The habit plane of martensite in steel depends on the carbon content. The internal structure due to the inhomogeneous shear also depends on carbon content and M_s temperature. The shear, *s*, shown, is the resultant of both the homogeneous and inhomogeneous shears. α' refers to the martensite phase.

boundaries or inclusions, are not particularly good nucleation sites for martensite. Instead, martensite usually prefers to nucleate at dislocations, or groups of dislocations, within the grain where it can utilize the *strain energy* of the dislocation as an aid to nucleation.

The mechanism by which martensite nucleates and grows is not fully understood. Assuming that the nuclei are in the form of thin plates of radius *a* and thickness *c*, equation (2.29) can be derived, where μ is the shear modulus

$$a^* = \frac{4\gamma_{\alpha'/\gamma}\mu s^2}{(\Delta G_V)^2} \tag{2.29}$$

of the austenite and *s* is the shear imposed on the austenite lattice. Thus, the critical nucleus size is very dependent upon the undercooling (proportional to ΔG_V) and the amount of the shear. In steels, $s \approx 0.2$ and $a/c \approx 30$ which gives a critical radius, a^*, of *ca.* 20—30 nm. Indeed, this critical nucleus size is large enough to exclude homogeneous nucleation of martensite.

The potency of a dislocation as a nucleation site depends on the size of its contribution to s. It is found from experiment that an increase in the dislocation density only slightly increases the number of martensite nucleation sites at the M_s (martensite start) temperature; the M_s is raised slightly. One reason for this is that the growth of martensite may be impeded if the dislocation density is too high. Since martensite growth is not a thermally activated process it can occur extremely rapidly, reaching speeds of *ca.* 1000 m s^{-1}. The mode of growth is not properly understood, except that to achieve the lattice structure of body-centred tetragonal (b.c.t.) iron, both a homogeneous shear and an inhomogeneous shear in the form of slip or twinning is needed. The mode of inhomogeneous shear adopted depends on the transformation temperature and carbon content. For carbon contents up to *ca.* 0.4 wt% the dominant mode is by slip and the growth morphology is then *lath*, i.e. elongated thin plates. Thus, lath martensite is that usually observed in steel weld metals. It grows in packets on $\{111\}_\gamma$ *habit planes*, with a *Kurdjumov–Sachs orientation relationship*, i.e.:

$(111)_\gamma // (011)_{\alpha'}$

$[10\bar{1}] // [\bar{1}1\bar{1}]_{\alpha'}$

Lath martensite, as observed in the electron microscope, has a very high density of dislocations, partly as a result of the inhomogeneous shear, but also because of the large plastic strains caused by this transformation. The high hardness of lath martensite is mainly due to this high dislocation density, but there is also some contribution to hardening by the carbon in solution.

The M_s temperature, which denotes the temperature at which martensite begins to form, does not mark the temperature at which the transformation goes to completion. Indeed, the reaction is not complete until the M_f (martensite finish) temperature is reached, *Figure 2.36*. Thus, for the alloy shown, $\Delta T \approx 200$ °C at the M_s and increases to ≈ 400 °C at M_f. This indicates

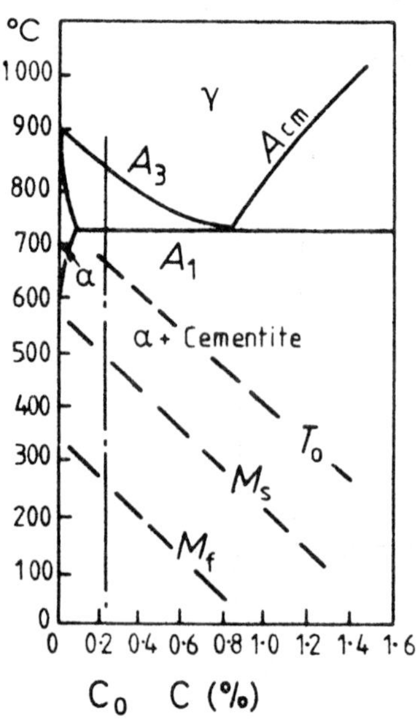

Figure 2.36 The M_s and M_f temperatures in steel as a function of carbon content. Compare with *Figure 2.28* in which $\Delta T = T_0 - M_s$ for this case. The alloy C_0 is typical of weld metal compositions

that critical martensite nuclei can only utilize a certain number of dislocation sites at the M_s, and that larger ΔT values are necessary if sites of less potency are to be utilized. It seems likely that the orientation of the dislocation or dislocation array is the key factor if the strain fields of the martensite nucleus and dislocation are to 'interact' efficiently. In low carbon lath martensites, the M_f is so high ($\approx$300 °C) that some tempering is likely to occur during cooling, which results in the formation of carbon atmospheres around dislocations and this may further increase the strength of martensite.

Some of the principles that govern the kinetics of solid–solid phase transformations, the C-curve behaviour of diffusional transformations, and the unique characteristics of martensite have been discussed. These principles help to explain phase transformations that occur in duplex stainless steel and transformable steel weld metals.

Transformations in duplex stainless steel welds

Problems of hot cracking in stainless steels of type 18% Cr/10% Ni can be avoided by using a weld metal that gives a *duplex* (ferrite + austenite) structure. The beneficial effect of this is that, because of the fine dispersion of phases and

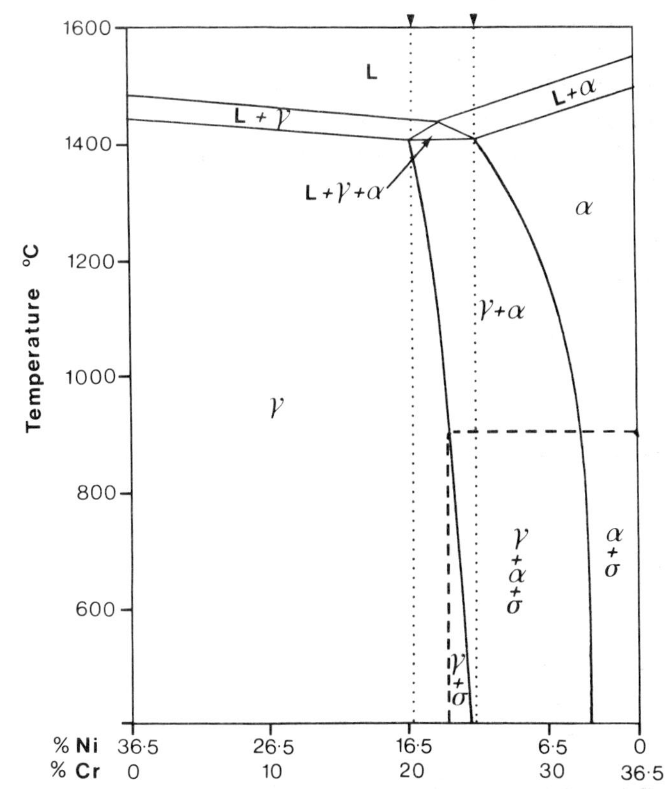

Figure 2.37 Vertical section of the ternary Fe–Cr–Ni phase diagram at 63.5 wt % Fe. The dotted lines define typical alloy ranges for duplex weld metals

the increase in phase boundary area, segregation of potentially dangerous impurities, such as P and S, can be reduced.

The duplex structure derives from the solidification characteristics of the alloy, shown in *Figure 2.37.* According to this vertical section of the Fe–Cr–Ni phase diagram at 63.5 wt % Fe, ternary alloys with compositions within the broken vertical lines, typical of these welds, should solidify as austenite plus δ-ferrite with either phase solidifying first depending on the (Ni–Cr) composition. Strictly, since weld solidification is initially *epitaxial,* the first

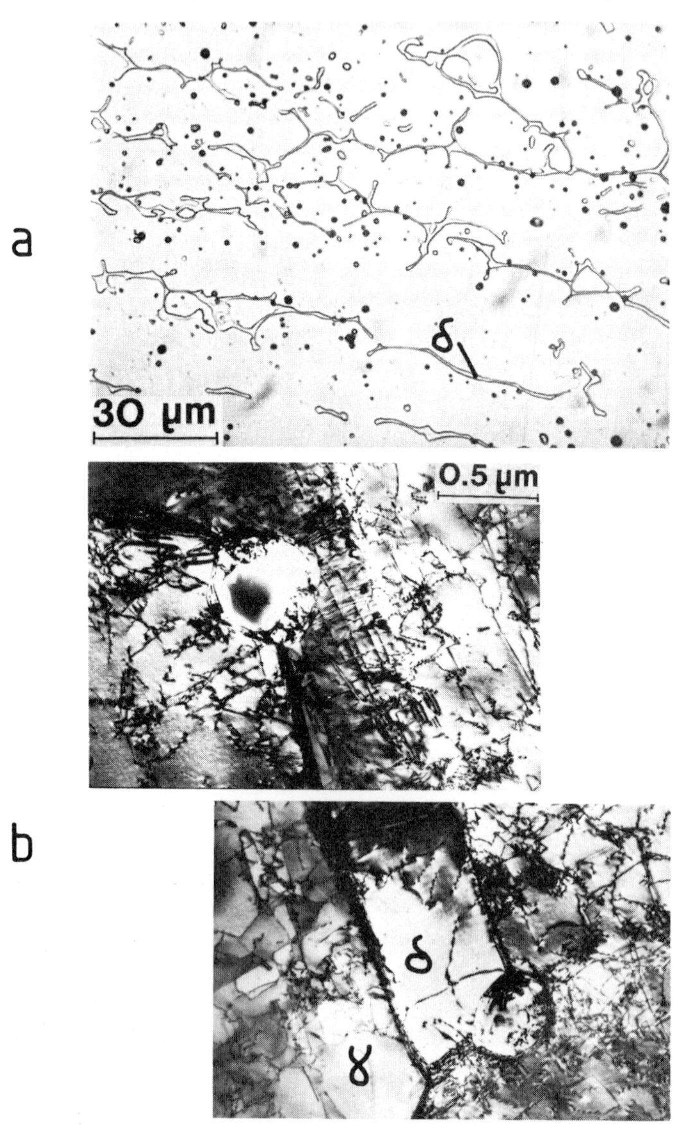

Figure 2.38 The duplex δ + γ weld metal within the composition range defined in *Figure 2.37;* (a) is an optical micrograph and (b) shows two transmission electron micrographs of the microstructure. Micrographs by courtesy of Dr Bengt Loberg, University of Luleå

phase formed is, in fact, determined not by the weld metal composition, but by the structure of the base metal at its melting point. Assuming this to be austenite, the duplex structure is expected to develop during the planar-to-cellular growth stage.

Cooling below the solidus increases the volume fraction of ferrite, as indicated by the phase diagram, However, owing to the fast cooling rates of welds some 6–10 vol% of δ-ferrite is retained at ambient temperature, depending on the cooling cycle. The microstructure at this stage is in the form of a very fine, often sub-micron interdendritic skeleton of ferrite in a matrix of austenite, as shown in *Figure 2.38*. Thin-foil X-ray microanalysis measurements[11] have shown that all phosphorus in the weld metal is contained in the δ-ferrite, with a maximum at the phase boundaries. Slag inclusions in the weld metal are fairly homogeneously dispersed between both phases with the majority located at phase boundaries. These tend to be rather complex, as may be expected (*see*, e.g., *Table 1.2*, p.5), and compositions of some of the inclusions studied are shown in *Table 2.2*. Sulphur is always contained in the Mn-based inclusions, which emphasizes the importance of Mn in these weld metals, both as an oxygen and sulphur scavenger.

TABLE 2.2 Chemical analysis of inclusions in 18% Cr/10% Ni manual metal arc weld metal (after ref. 11)

Inclusion	*Location*	*Elements present*	*Main constituent*
1	Austenite	Si Cr Ti FeS (O)	Si
2	Austenite	Fe S Cr (O)	S
3	Austenite	Si Ti Cr Al MnS (O)	Si
4	Ferrite	Mn S Si Ti Cr (O)	Cr
5	Phase boundary	Mn S Si Ti Cr (O)	S
6	Phase boundary	Mn S Si Ti Cr Cu (O)	S
7	Phase boundary	Si Ti Cr Al Mn (O)	Cr
8	Phase boundary	Si Cr (O)	Si and Cr
9	Phase boundary	Fe Mn Cr (O)	Cr
10	Phase boundary	Si Cr Ti Mn S (O)	Si
11	Phase boundary	Mn S Si Cr (O)	Cr
12	Phase boundary	Mn S Si Ti Cr Cu (O)	Cr, Mn and S
13	Phase boundary	Si (O)	Si
14	Phase boundary	Mn S Si Cr Cu (O)	Si, Cr, Mn, S

The amount of ferrite that remains at ambient temperature is clearly important in these welds, and it can be predicted with the aid of a *Schaeffler diagram, Figure 2.39* (*a*). This plots volume fraction of phases on the basis of *Ni- and Cr-equivalents,* as estimated from the formulae given in the diagram. If nitrogen is present in the weld metal, which is a strong austenite stabilizer, a *De Long diagram* has to be used instead to estimate the residual ferrite, *Figure 2.39* (*b*). There should not, in fact, be more than *ca.* 10% ferrite if the good corrosion properties of the weld metal are to be retained.

Transmission electron microscopy shows that the phase boundaries are high angle (incoherent) so that growth of the austenite phase at the expense of ferrite during cooling has to be a diffusion-controlled process. The problem is that as the austenite phase absorbs the ferrite large adjustments in composition

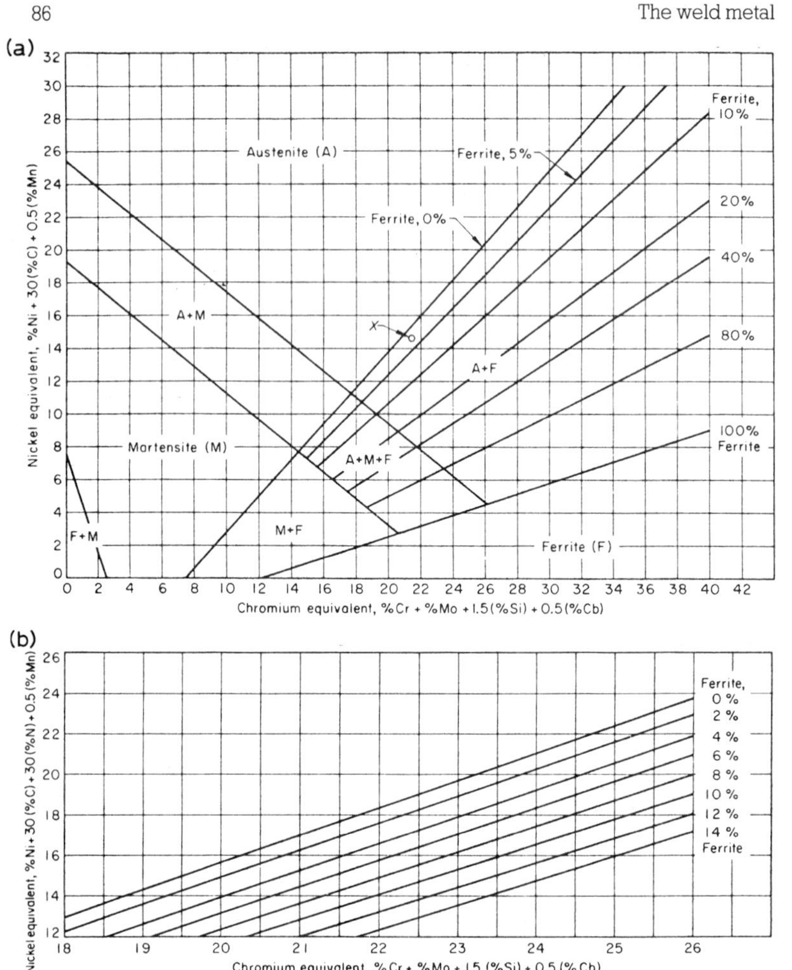

Figure 2.39 (a) A Schaeffler diagram. (b) A DeLong diagram. Taken from Taylor, L. (ed.), Welding and brazing, Vol. 6, *Metals Handbook*, 8th edn, American Society for Metals, 1971

have to be accommodated. Consider the hypothetical α/γ interface in *Figure 2.40* (*a*). Assume, for the sake of simplicity, that equilibrium is maintained at the interface and a compositional change that involves only one solute is required. If a unit area of the interface moves forward an increment, dx, the quantity of solute to be removed is dx ($C_i^\alpha - C_i^\gamma$), where C_i^α and C_i^γ refer to the solubilities of solute at the ferrite–austenite interface. Thus, the velocity R_i of the interface of this diffusion-controlled reaction is given by equation (2.30),

$$R_i = \frac{D}{(C_i^\alpha - C_i^\gamma)} \cdot \frac{\partial C}{\partial x} \tag{2.30}$$

where D refers to the diffusivity of the solute in the ferrite. As the austenite phase expands at the expense of ferrite the diffusion distance of solute into the ferrite is extended (*Figure 2.40* (*b*)) since C^α reaches its limit of solubility. The effective diffusion distance, $\bar{x}$, at any given time can be obtained from equation

(2.31). Now, since $\bar{x}$ increases with time, while $(C_i^\alpha - C^\alpha)$ remains constant, the

$$\frac{\partial C}{\partial x} = \frac{C_i^\alpha - C^\alpha}{\bar{x}} \tag{2.31}$$

growth rate of austenite falls with time and equation (2.32) can be derived[3].

$$\bar{x} \approx (Dt)^{1/2} \tag{2.32}$$

This indicates that although the austenite growth rate is initially high it falls off rapidly with time, particularly under the continuous cooling conditions of the weld. Note that for solutes that segregate strongly to boundaries, $(C_i^\alpha - C^\alpha)$ increases, thereby effectively exerting a 'drag' on the movement of the interface.

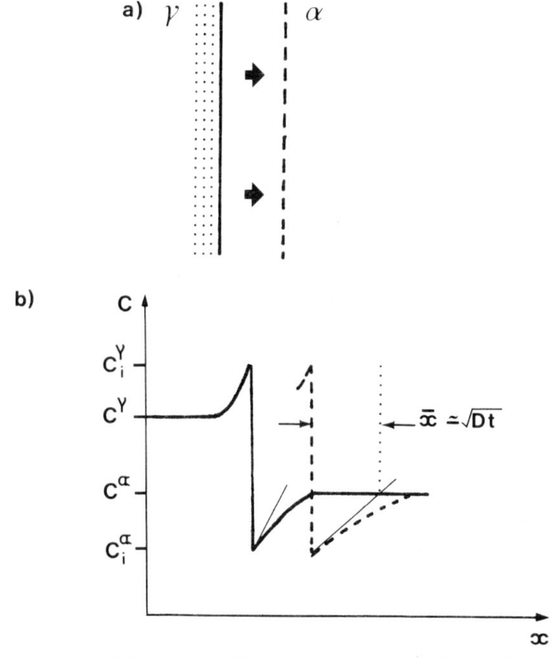

Figure 2.40 (a) Schematic diagram showing the diffusion profiles ahead of an expanding phase boundary. The effective diffusion distance, $\bar{x} \sim (Dt)^{1/2}$ is indicated in (b)

The concentration profiles in a duplex 19% Cr/12% Ni weld metal were recently obtained[12] using thin-foil X-ray microanalysis and the results are shown in *Figure 2.41*. The spacial resolution of this microanalysis technique is *ca.* 50 nm for quantitative measurements, i.e. approximately 100 times better than conventional (SEM-based) probe microanalysis, which allows reasonable measurements to be made even at the phase boundaries. The schematic concentration profile in *Figure 2.40 (b)* can be compared with those shown in *Figure 2.41,* which shows that Ni, Cr and Mo all require significant redistribution at the interface between the γ- and α-phases. It was found that Nb, while

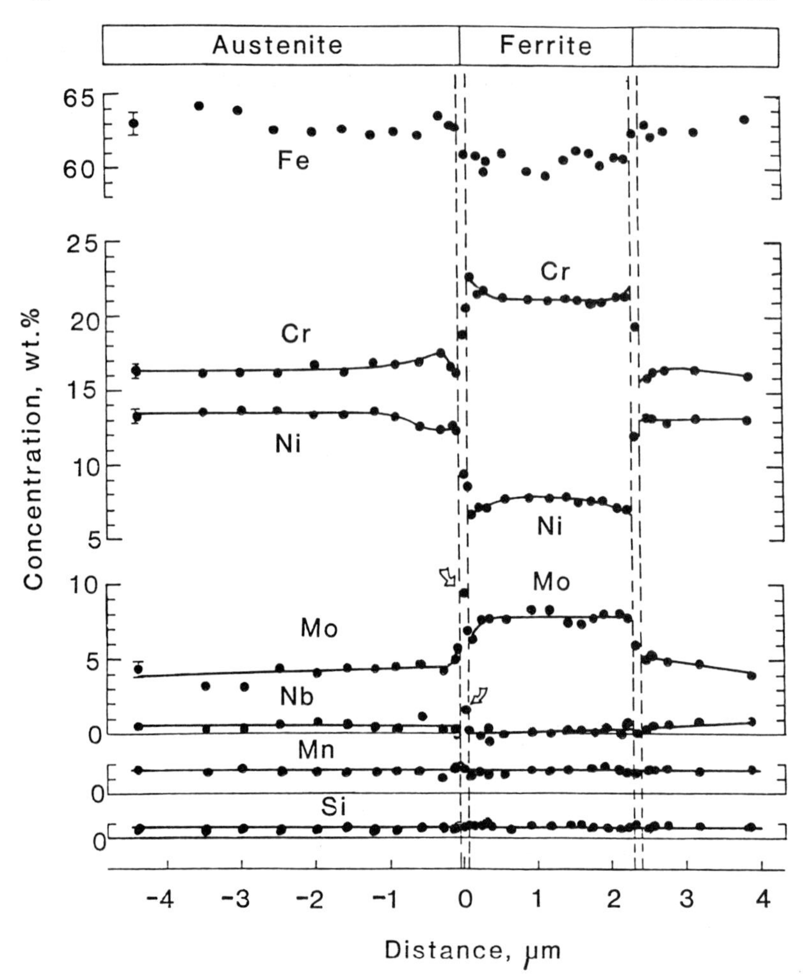

Figure 2.41 Measured composition profiles across the austenite–ferrite phase boundaries in a duplex weld metal. The measurements are made using an energy dispersive X-ray spectrometer attachment to a 200 kV scanning-transmission electron microscope. After Kyröläinen and Porter [12]

present in relatively small amounts, appears to segregate strongly to the phase boundary. In fact, the partitioning coefficient, k, of Nb in steel (*see* equation (2.11)) is known to be small, which suggests that this element is expected to segregate strongly at boundaries. The high concentrations of Cr and Mo at the phase boundary are thought to assist in σ-phase formation when reheating these welds[12].

Transformations in carbon and low alloy steel welds

The discussion herein initially concerns a low carbon steel of, nominally, 0.15—0.2 wt % C; the effect of possible impurities and alloying elements on the transformation characteristics is discussed later.

Following solidification, the steel has the characteristic cellular– dendritic structure of cast metal, consisting of both rather coarse, columnar austenitic grains which curve into the weld centre line and a fine cellular network within the grains. Since the cell boundaries are *small angle* consisting of dislocation networks, they are not likely to be very potent nucleation sites for ferrite when

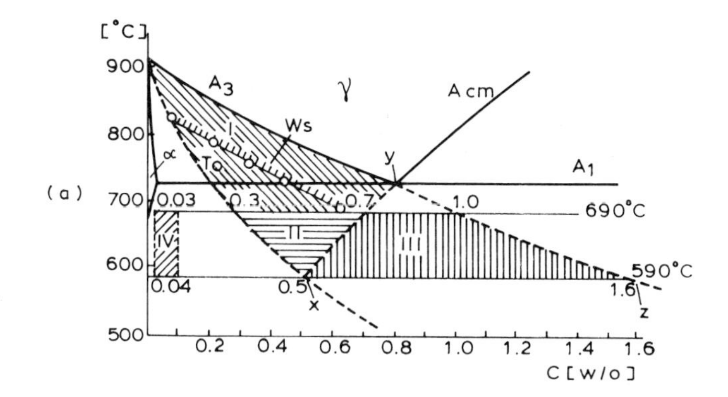

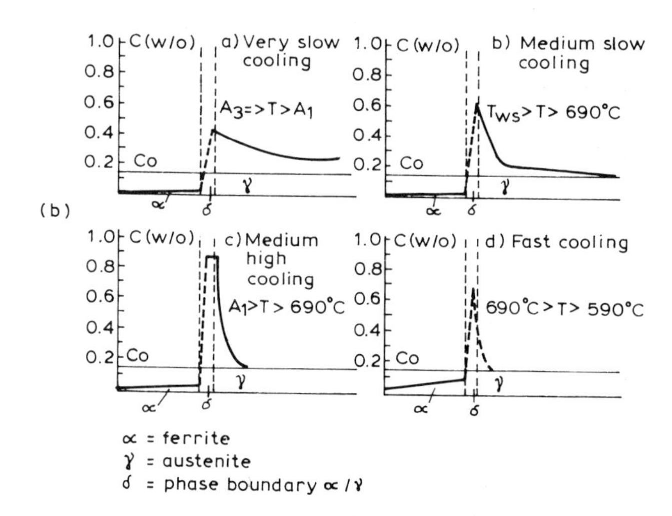

Figure 2.42 (a) The quasi-equilibrium diagram of the system Fe–Fe_3C, showing the various transformation regions (*see* text). (b) Carbon profiles at the γ/α interface for various cooling rates. After Räsänen and Tenkula[13]

ooling below the A_3 temperature, and initially ferrite is expected to form at ne large angle boundaries between the columnar austenitic grains. However, s discussed earlier (*see* Kinetics of phase transformations, pp. 75–83), the ansformation product chosen is dependent upon the rate of cooling of the eld metal. It is useful, therefore, to discuss the transformation product as a

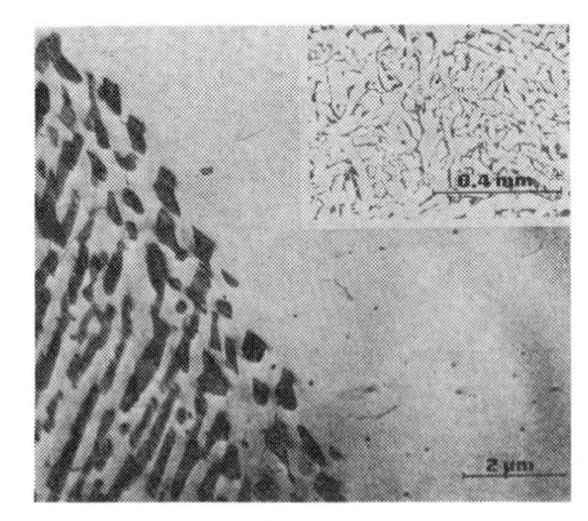

(a) ferrite and pearlite

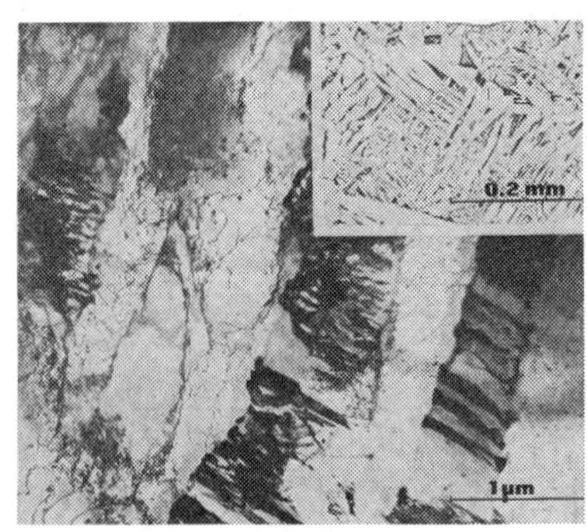

(b) Widmanstätten side plates and pearlite

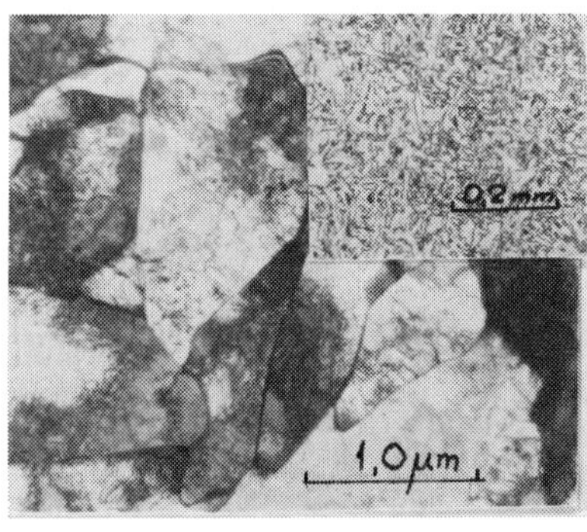

(c) acicular ferrite

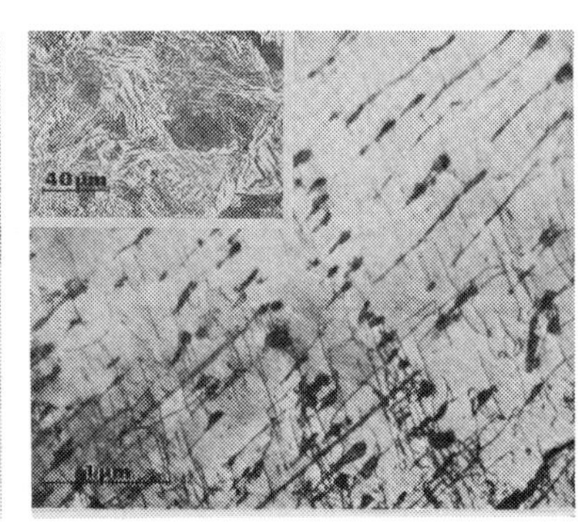

(d) periodic pearlite

Figure 2.43 (a)—(d) Examples of microstructures observed in weld metals after cooling. All micrographs except (c) are after Räsänen and Tenkula[13]. In all cases the figures are transmission electron micrographs (TEM) with light optical inserts

function of cooling rate, and with reference to a metastable $Fe-Fe_3C$ phas diagram, *Figure 2.42 (a)*. Thus, the arguments developed by Räsänen an Tenkula[13] are closely followed.

Slow cooling rate (A) Ferrite nucleates at high angle grain corners an boundaries at a small undercooling below A_3, and grows into the austenit behind a planar front. Redistribution of carbon occurs efficiently, with a solut profile as in *Figure 2.42 (b)*, top left. Indeed, at these high temperature carbon diffusion occurs rapidly. Using equation (2.32) and assuming diffusivity of carbon in austenite of *ca.* $10^{-11}\,m^2\,s^{-1}$ at the A_3 temperature, the a solute build-up, $\bar{x}$, of 2 μm occurs in about a tenth of a second — a short tim even by weld cooling rate standards. The carbon content C^{γ} at the interface then within area I of *Figure 2.42 (a)*. As the temperature decreases, carbo continues to diffuse into the austenite with the growth of 'block *allotriomorphs* of ferrite, until the temperature falls below A_1. At this stag the austenite regions that are enriched with carbon to an amount within th

(e) upper bainite plus cementite

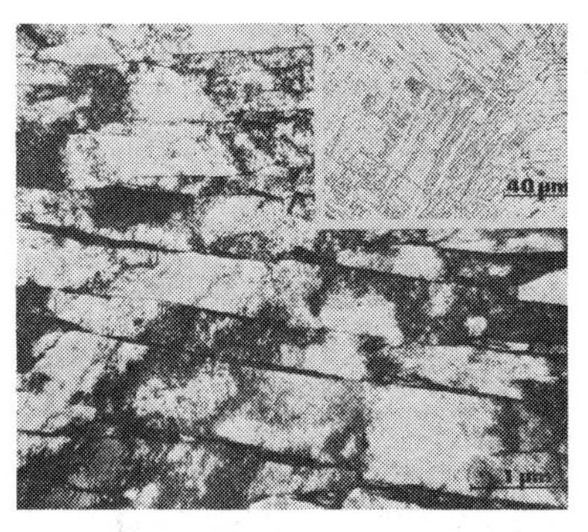

(f) upper bainite and retained austenite.

(g) lower bainite

(h) lath martensite

Figure 2.43 (e)—(h) Examples of microstructures observed in weld metals after cooling. After Räsänen and Tenkula[13]. In all cases except (h) the figures are TEM with light optical inserts, (h) is TEM only

eutectoid triangle xyz (*Figure 2.42* (*a*)), transform into *pearlite,* so that a mixed structure of ferrite and pearlite is obtained, *Figure 2.43* (*a*).

Medium–slow cooling rate (*B*) The undercooling below A_3 is larger than for case (*A*) and, while ferrite nucleates and grows efficiently from the grain boundaries and moves into the austenite behind a planar front, the carbon redistribution is not as efficient as with slower cooling (*Figure 2.42* (*b*) top right) and the solute gradient into the austenite is steeper. This form of ferrite, which quickly decorates all grain boundaries, is referred to as *proeutectoid ferrite.* It can be distinguished from blocky ferrite of (*A*) by its mode of growth, which occurs behind an essentially incoherent (large-angle) phase boundary. Blocky ferrite, on the other hand, expands behind both semi-coherent and incoherent boundaries. Growth by the movement of an incoherent boundary requires, in fact, a higher ΔG driving force, i.e. higher undercooling, ΔT, than does growth by movement of the lower energy semi-coherent boundary, at least when long range diffusion is involved.

When the austenite undercools below the line W_s (denoting the boundary that divides proeutectoid ferrite and Widmanstätten side-plate formation) in *Figure 2.42* (*a*), only separate ferrite needles can break through the carbon–solute barrier. These needles, known as *Widmanstätten side plates* have the Kurdjumov–Sachs orientation relationship with the austenite. Growth of the side plates is rapid since solute is efficiently redistributed to the sides of the growing tips, thus avoiding solute pile-up problems. On further cooling below A_1 the carbon-enriched regions between the side plates transform to cementite or pearlite. The resulting microstructure is illustrated in *Figure 2.43* (*b*).

Another reaction that can occur in weld metals at undercooling near to A_1 is the formation of *acicular ferrite*. This is a more equiaxed product and its growth is facilitated by utilizing a high density of nucleation sites, including inclusions within grains. The possible important role of inclusions in helping to nucleate acicular ferrite is taken up later (*see* pp. 96—98). Acicular ferrite is also able to form if the proeutectoid reaction is suppressed, e.g. by carbide formation at austenite–ferrite boundaries. This phase is illustrated in *Figure 2.43* (*c*). It appears that strong carbide forming elements, such as Mo and Cr, are particularly effective in limiting proeutectoid ferrite growth and thus enhance the acicular ferrite transformation, which gives improved toughness to these welds.

Medium–high cooling rate (*C*) In this case, ΔT is sufficient for the transformation to occur below A_1, and the initial growth rate of ferrite is so rapid that the carbon concentration at the interface immediately enters the region of the *xyz* triangle in *Figure 2.42* (*a*), assuming a concentration at the austenite–ferrite interface, as shown in *Figure 2.42* (*b*), bottom left. The austenite then becomes unstable with respect to both ferrite and cementite, so that when a small amount of Fe_3C is precipitated, ferrite continues to grow until a carbon barrier is again formed. This transformation is known as a *periodic pearlite reaction,* and is illustrated in *Figure 2.43* (*d*).

Fast cooling rate (*D*) In this case, it is assumed that ΔT depresses the transformation temperature to below *ca.* 690 °C, which means that carbon does not have time to diffuse into the austenite as the grain boundary ferrite forms. Instead, it concentrates and redistributes at the phase boundary (*Figure 2.42* (*b*) bottom right) and the austenite transforms to ferrite either by a partial *massive* mechanism or by the formation of *upper bainite*. A massive transformation occurs by the movement of a high angle phase boundary, such that atoms move from the parent to the product lattice by short range diffusion across the boundary. This implies that the carbon distribution in both phases is the same. In the case of upper bainite, however, the ferrite nucleates at grain boundaries and grows rapidly into the grain behind a low energy semi-coherent interface. It then develops in the form of ferrite fingers with Fe_3C precipitating between them. As the carbon content at the moving phase boundary increases it eventually attains a level at which cementite is able to nucleate and grow. At the higher temperatures of formation, upper bainite resembles finely spaced Widmanstätten side-plates. Examples of this microstructure are illustrated in *Figures 2.43* (*e*) and (*f*).

'ery fast cooling rate (*E*) In this case, ΔT is such that transformation occurs in he region of *ca.* 500 °C, i.e. the M_s temperature, and transformation is then a liffusionless process to produce *lath martensite* (growing independently of ;rain boundaries, as discussed on pp. 80–83). Alternatively, at temperatures ust above the M_s, *lower bainite* may grow if carbon diffusion at the austenite ;rain boundary occurs sufficiently to allow laths of bainite to nucleate and ;row.

Lower bainite is essentially an interface controlled reaction in which very hort range diffusion of carbon occurs to form carbides with an orientation elationship to the ferrite, concurrent with a thickening and lengthening of the lates. At these low temperatures of growth, carbon diffusion is so slow that he ferrite plates seem to thicken more easily by the repeated precipitation of ementite at the interface. This, in effect, removes any excess of carbon in the ustenite and thus allows the interface to advance. The final morphology of ower bainite is therefore more lath-like than needle-shaped.

At very high cooling rates, however, there is no time for carbon diffusion nd martensite forms instead. This low carbon lath martensite is believed to ontain very thin regions of retained austenite between laths, or packets of aths, evidently due to some limited carbon enrichment[14]. Examples of lower ainite and lath martensite are shown in *Figures 2.43* (*g*) and (*h*).

Additional factors that can affect these various transformation products re the austenite grain size (as discussed in the previous sections) and the resence of impurities or alloying additions; the latter is considered below.

Role of alloying in transformation kinetics

In the production of structural steels, the transformation C-curve haracteristics can effectively be controlled by careful choice of alloying lements. In weld metals, however, the composition is decided by factors other han the composition of the filler wire, such as dilution, fluxes, gases, etc., as liscussed in Chapter 1. The final composition of the deposit typically lies etween that of the filler metal and the base metal. For some good examples of uch alloying effects in C–Mn steel weld deposits, *see* refs 15 and 16.

Some elements help to stabilize the austenite (austenite formers) and thers stabilize ferrite (ferrite formers). Included in the first category are C, N, Mn, Ni, Cu and Zn, given in decreasing order of effectiveness. The latter ategory includes Zr, Ti, P, V, Nb, W, Mo, Al, Si and Cr. Of the elements nentioned, C, N, Mn, Mo and Si are all typically present in steel weld metals. However, the problem of how these various alloying elements affect the $\gamma \rightarrow \alpha$ ransformation — and hence C-curve behaviour — in steels is complicated by he tendency of some elements to exhibit partitioning at the γ/α interface (for a ood discussion of this *see* Honeycombe[17]).

An example of the large effect of carbon on the C-curve behaviour of a C–Mn steel is shown in *Figure 2.44*, where an increase from 0.19 wt % C to .28 wt % C results in a much broader bainitic field and a wider martensitic eld. The role of Mo is complicated, but it is claimed to enhance transformaon to acicular ferrite at the expense of proeutectoid ferrite. The reason is that by forming carbides in the austenite it can inhibit movement of the extensive planar front of the proeutectoid phase by pinning or dragging effects. On the other hand, at lower transformation temperatures, carbides cannot suppress acicular ferrite growth because of the higher density of nucleation sites available. Cr, another strong carbide former, is claimed to have a similar effect to that of Mo.

The effects of alloying elements present in the *microalloyed steels,* e.g. Nb, V, Al and Ti, have also to be accounted for, since through dilution in high energy welding they are typically present in weld deposits in amounts up to

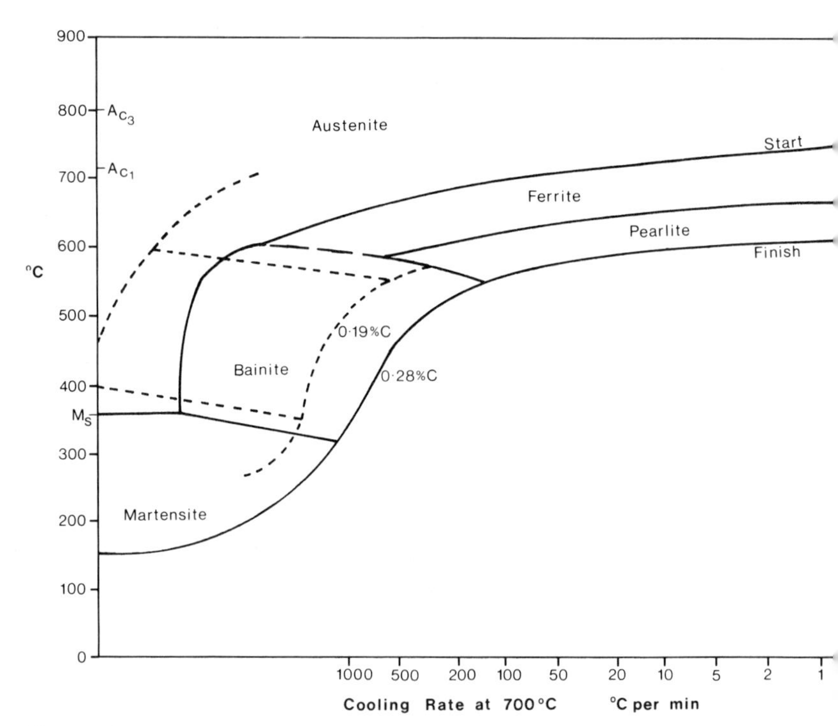

Figure 2.44 CCT curves of two plain C–Mn steels with 0.19 wt % C (dotted) and 0.28 wt % C (full lines). Note the effect of a relatively small increase in C-content on the positions of the martensite and bainite fields. Adapted from Atkins, M., *Atlas of Continuous Cooling Transformation Diagrams for Engineering Steels,* British Steel Corporation, 1977

ca. 60—70 % of the base metal composition, i.e. 0.02 to 0.03 wt % in the weld deposit in the case of Nb, and even higher for V. Of the microalloying elements mentioned, Nb and V thus deserve special consideration.

Niobium combines readily with carbon to precipitate as NbC at temperatures below *ca.* 1000 °C. At the high cooling rates normally associated with welds there may be insufficient time for carbides to grow during cooling, although they are often observed to form during reheating, e.g. during stress relief annealing or after subsequent weld runs. An example of NbC precipitation on dislocations in a reheated weld metal is shown in *Figure 2.45 (a)*. Stress relief annealing treatments are typically carried out at 550—650 °C for

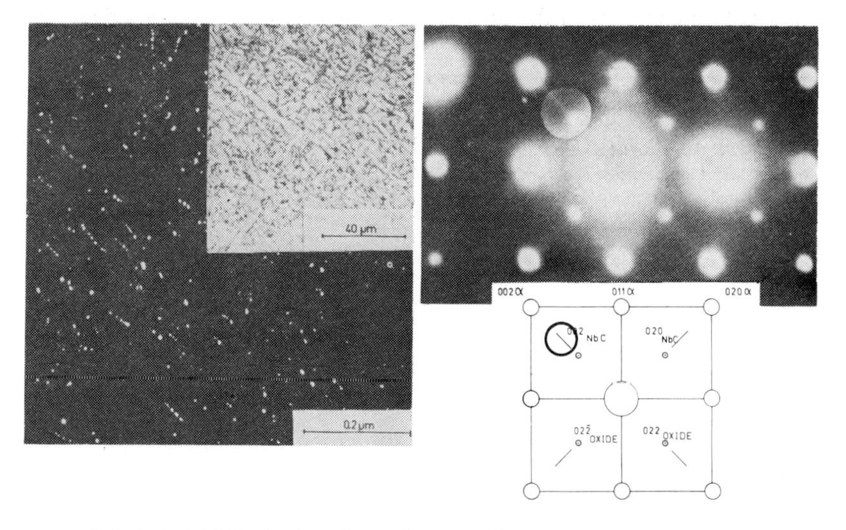

(a) dark field TEM using diffraction condition shown

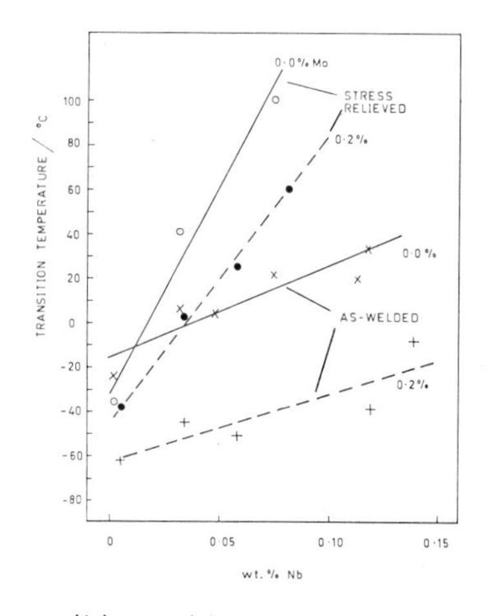

(b) transition temperatures

Figure 2.45 (a) NbC precipitation in a stress-relieved weld metal. (b) The increase in ductile–brittle transition temperature in Nb/Mo-steel weld metals due to NbC precipitation. After Bosansky, J., Porter, D.A., Åström, H. and Easterling, K.E., The effect of stress annealing treatments on the structure of high heat-input welds containing Nb and Mo, *Scandinavian Journal of Metals*, **6,** 125, 1977

an hour, and in microalloyed steel welds in particular this can cause precipitation which results in matrix hardening and a decrease in toughness, *Figure 2.45* (*b*). Nb is thought to segregate to the austenite–ferrite interface (during cooling of the weld, *see Figure 2.41*) where it tends to suppress the growth of proeutectoid ferrite. On the other hand it cannot suppress bainite formation and the presence of Nb in solution in austenite is generally considered to enhance the lower transformation products, particularly bainite.

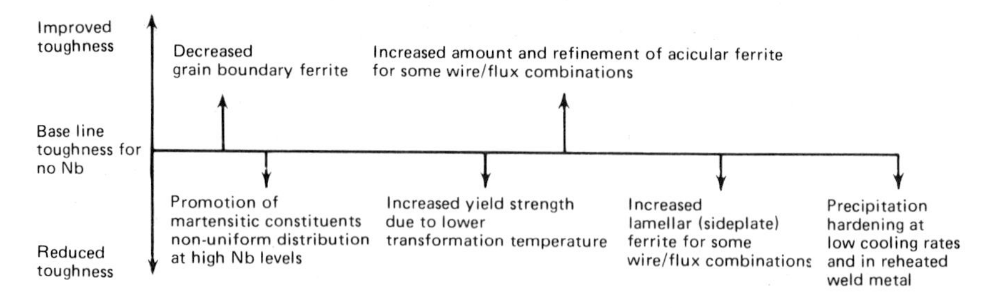

Figure 2.46 The effect of niobium on weld metal toughness. After Dolby[18]

Dolby[18] has attempted to tabulate the positive and negative effects of Nb on weld microstructure and toughness, as shown in *Figure 2.46*.

Vanadium combines with carbon and/or nitrogen to precipitate as V(CN) at temperatures below *ca.* 800 °C. As with Nb, however, precipitation is rarely seen in conjunction with cooling of the weld metal, but is often observed on reheating. A positive effect is that V tends to promote the acicular ferrite

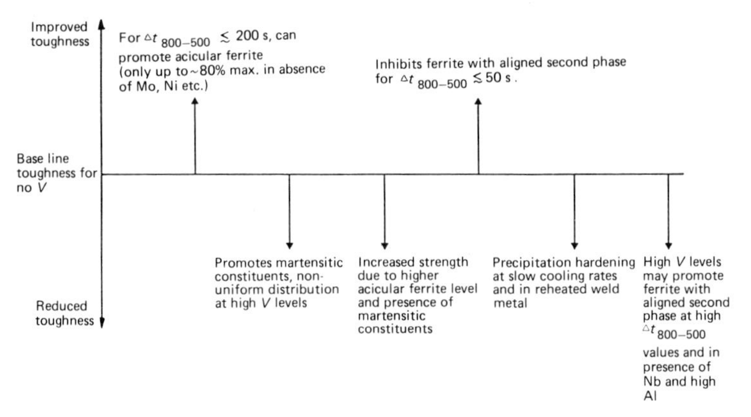

Figure 2.47 The effect of vanadium on weld metal toughness. After Dolby, R.E., *Review of Work on the Influence of Vanadium on the Microstructure and Toughness of Ferritic Weld Metal*, International Institute of Welding, Document No IX–1213–1981, 1981

transformation, due presumably to its inhibiting effect in the form of V(CN) precipitation on proeutectoid ferrite growth. As with Nb, however, V(CN) precipitation has the effect of increasing the yield strength of the weld metal, particularly if reheating is necessary, and this can be detrimental to toughness. *Figure 2.47* shows the relative merits of V on the microstructure and toughness of weld deposits.

Role of slag inclusions in transformation kinetics

It is shown in Chapter 1 (*see* pp. 4–17) that steel weld deposits are likely to have a fairly high volume fraction of inclusions compared with normal steels, epending on such factors as flux basicity, deoxidation elements used, etc. The uestion now arises as to how effective these inclusions can be as possible ucleation sites for ferrite.

As discussed earlier in this chapter (pp. 75–80), the effectiveness of a ucleation site is controlled by thermodynamic considerations, particularly urface and strain energies, and the density of the sites. Assuming that, e.g., in igh rutile flux MMA welds the density of inclusions is high, and that ferrite ucleation and growth from grain boundaries has been suppressed (e.g. by arbide formation), it is of interest to estimate the potency of slag inclusions as function of their species and sizes.

Inclusions in welds are usually of very complicated composition, often omprising a number of elements and even phases (*see*, e.g., *Table 2.2*). They re, however, usually based on mixed (MnAl) silicates and oxides, and are thus ssumed to exhibit a fully incoherent (high energy) interface with the matrix.

Consider the formation of a ferrite nucleus in the shape of a spherical egment at the inclusion–matrix interface, as illustrated in *Figure 2.48*. It can e shown that the total change in free energy associated with nucleation is given y equation (2.33), where $\gamma_{\gamma/\alpha}$ refers to the ferrite–austenite interfacial energy

$$\Delta G = \Delta G_V + A\gamma_{\gamma/\alpha} \tag{2.33}$$

nd equation (2.34) gives A, the area of the nucleus after subtracting the

$$A = 4\pi r^2 - 2\pi r^2 (1 - \cos\theta) \tag{2.34}$$

ucleus–inclusion interface. It is assumed that the inclusion–ferrite and errite–austenite interfacial energies are similar. Differentiating equation 2.33) with respect to the nucleus radius and equating to zero enables the effect

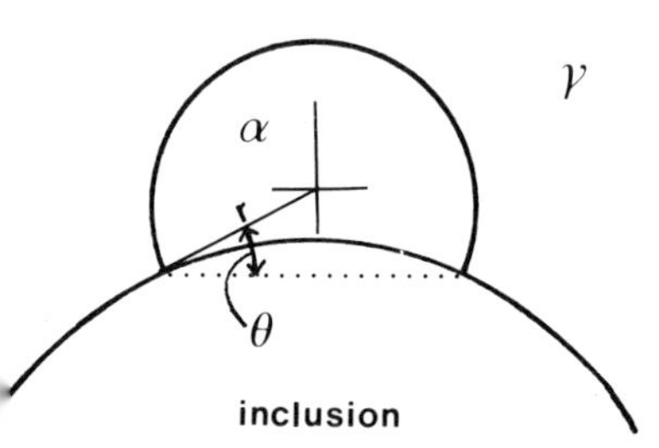

Figure 2.48 Nucleation of a spherical nucleus of ferrite at an inclusion in the austenite

f inclusion *size* on the size of the nucleation barrier, i.e. the effectiveness of e inclusion as a nucleation site, to be evaluated. This has been done by Ricks, arritte and Howell[19], and their results are shown in *Figure 2.49*, in which the nergy barrier to nucleation (normalized with respect to the homogeneous ucleation barrier, ΔG^*_h) is plotted as a function of inclusion size. As expected, rain boundaries are always more favourable sites for ferrite nucleation than clusions. However, assuming that ferrite growth from grain boundaries is indered in some way, it appears that inclusions are still not very good sites for errite nucleation unless they attain a certain minimum size of *ca.* ¼—½ μm.

In fact, many slag inclusions in weld metals are of this size range and on thi basis it can be concluded that they may contribute to ferrite formation Furthermore, for a uniform dispersion of inclusions of this size range (o larger) all act as nucleation sites which tends to produce a relatively equiaxe acicular ferrite microstructure in the weld deposit.

Another factor which may contribute to the potency of inclusions as site for ferrite nucleation and recrystallization concerns the stresses and strains tha arise from the difference in thermal contraction between inclusions an austenite. It can be shown from elasticity theory that the maximum stres

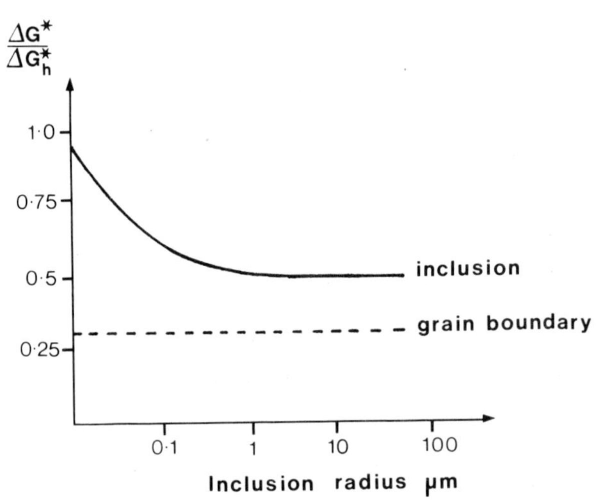

Figure 2.49 Plot of a normalized energy barrier to nucleation of ferrite at an inclusion and at a grain boundary. After Ricks, Barritte and Howell[19]

generated at the interface of a spherical inclusion during cooling is given b equation (2.35)[20], where K^i and K refer to the bulk moduli of the inclusion ar

$$\sigma_{max} = \left(\frac{6K^i\mu\Delta T}{2K^i + K}\right)(\alpha - \alpha^i) \qquad (2.3$$

matrix, α^i and α are the respective coefficients of expansion, ΔT is th temperature range considered and μ refers to the shear modulus of th austenite. Considering the case of silica inclusions in austenite, in whic $K^i/K \approx 1/2$, $\alpha = 18 \times 10^{-5}$ °C^{-1}, $\alpha^i = 0.05 \times 10^{-5}$ °C^{-1}, and assuming a ΔT 600 °C, the stress at the inclusion is estimated to be about $\mu/6$. This is a ve large stress indeed and exceeds that needed to nucleate new dislocations at th interface (*ca.* $\mu/40$). Thus, the large stresses and strains that result fro difference in contraction between silica inclusions and austenite may aid th formation of ferrite by, e.g. helping to create the right conditions for ferri growth with the help of glissile dislocations nucleated at these inclusions.

Predicting the microstructure and properties of weld metals

Those features of weld solidification that are most likely to influence the final microstructure of the weld metal after cooling to ambient temperatures have been presented above. The overall picture is complicated by a number of interacting factors which include:

1. The welding process itself which determines the weld pool size and geometry.
2. The final composition of the melt as influenced by the filler wire, the base metal, fluxes, gases, moisture in the air, etc., and its effect on constitutional supercooling and segregation.
3. The speed of welding and its effect on solidification speeds, crystal morphology and segregation.
4. The weld thermal cycle and its influence on microstructural coarseness and type of transformation product produced during cooling.
5. The effect of weld metal composition, particularly from dilution in high energy welding of microalloyed steels, on precipitation reactions, and especially during reheating or in multi-run welds.

It appears to be unrealistic to attempt to develop CCT diagrams specific to weld metal composition and thermal history. On the other hand, it is useful and informative to express the influence of the various features of welding, as listed above, in a schematic CCT diagram which shows, e.g., the tendency for the C-curves to move to longer or shorter times or the introducion of shape or size changes of the transformation fields, and this is illustrated in *Figure 2.50.*

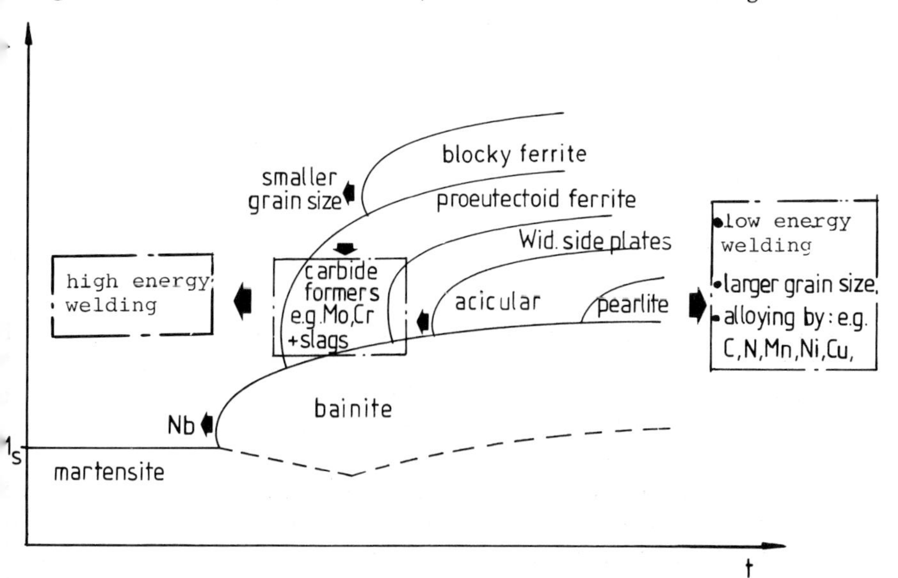

Figure 2.50 Schematic CCT diagram for steel weld metal, summarizing the possible effect of microstructure and alloying on the transformation products for a given weld cooling time

Arrows that point left in the diagram denote movement of C-curves to shorter transformation times, and arrows to the right indicate the opposite effect. Thus, austenite stabilizers (e.g. C, N, Mn, Ni, Cu), tend to inhibit transformation, pulling the C-curves towards longer times to transformation. Strong carbide or nitride forming elements (e.g. Mo, Cr, Nb, V, Ti, Al) however, tend to suppress blocky and proeutectoid ferrite, but not acicular ferrite or bainite. Indeed, Nb in particular tends to enhance bainite formation. Slag inclusions, particularly if present in sufficient number and size, also tend to promote the nucleation of acicular ferrite.

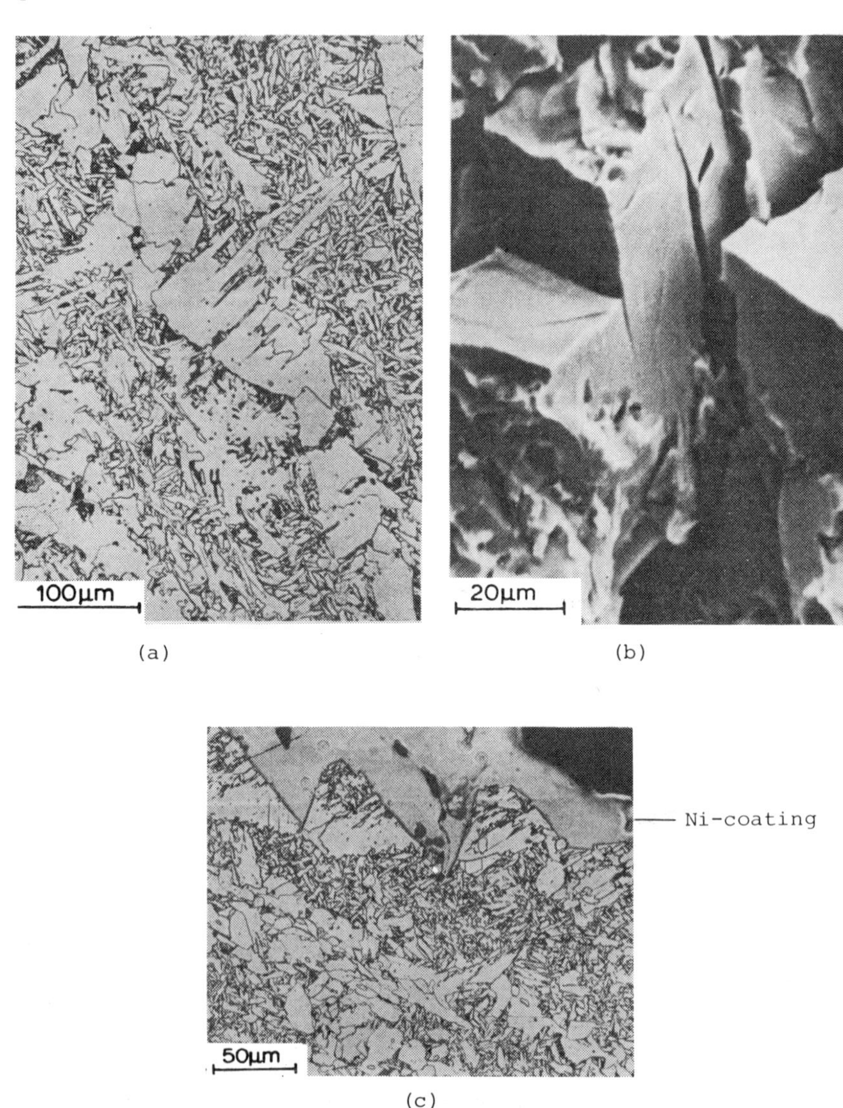

Figure 2.51 The brittle fracture of proeutectoid ferrite (light phase in (a) and (c)) in a steel weld metal. (b) is a SEM micrograph from a broken Charpy specimen, and (c) is a light micrograph of the fracture surface. After Billy *et al.*[21]

During reheating, or possibly in multi-run welds, microalloying elements such as Nb, V, Ti, etc, may cause grain hardening and reduced toughness because of precipitation reactions.

The strength and hardness of welds is generally high. This is due to the abundance of impurities and alloying elements which, in combination with rapid cooling rates, tend to promote the lower temperatures of transformation. Many of these lower transformation products, particularly bainite and martensite, contain a very high density of dislocations, and this, together with

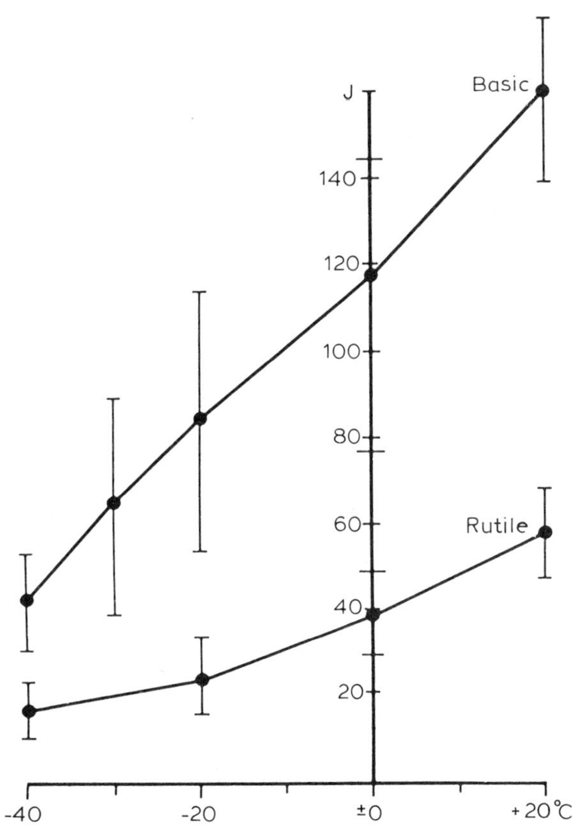

Figure 2.52 Influence of type of electrode coating on Charpy V impact energies of weld metal. After Almqvist *et al.* (ref. 5 in Chapter 1)

solution hardening, causes high hardness and strength in weld metals. An additional factor which probably contributes to weld hardness is plastic deformation due to residual stresses, and this increases the dislocation density in all the microstructural constituents.

The impact properties and toughness of steel welds are generally not high. This is discussed in more detail in Chapter 4, and is evidently due to the inhomogeneity of weld microstructure which results from its relatively coarse columnar solidification structure and from segregation. Hence, a mixture of transformation products is obtained in which proeutectoid ferrite often decorates columnar, austenite grain boundaries. Together with the high solute concentration at the ferrite–austenite interface this can cause carbide precipitation. Impact stresses tend to rupture these carbides to produce brittle cleavage fracture of the proeutectoid ferrite[21], a phenomenon illustrated in *Figure 2.51.* The weld centre line at which, e.g., the Charpy V-notch is normally located, has the highest segregation due to constitutional supercooling effects. The weld metal toughness is thus likely to be least in this location, particularly for the case of high speed welding.

The importance of slag basicity on notch toughness has been noted earlier (*see Figure 1.3*). For example, there is a considerable difference in impact values between high efficiency electrodes of the fully basic or the rutile type, and this is shown in *Figure 2.52.* The basic weld metal exhibits a rather steep transition curve and a wide scatter band with high upper shelf values. The rutile weld metal gives less scatter and since the upper shelf values are low, the transition temperature range is less well defined. The reasons for these differences are found in the distribution and form of the slag inclusions. The basic weld metal contains predominantly small ellipsoidal inclusions, mainly silicates, which tend to be evenly distributed. The rutile weld metal, on the other hand, typically contains a much wider range of inclusion sizes and a higher volume fraction than the basic metal.

Many of the complex variables that control the thermal history and composition of weld metals are not present in the case of the *heat-affected zone* of the base metal. The possibilities of making reasonable quantitative predictions of microstructure and properties are thus much improved. In the next chapter this part of the welded joint is considered in more detail.

References

1. Davies G.J. and Garland, J.G., Solidification structures and properties of fusion welds, *International Metals Review,* **20,** 83, 1975

2. Savage, W.F. and Aaronson, A.H., Preferred orientation in the weld fusion zone, *Welding Journal,* **45,** 85, 1966

3. Porter, D.A. and Easterling, K.E., *Phase Transformations in Metals and Alloys,* Van Nostrand Reinhold, p. 198, 1981

4. Savage, W.F., Solidification, segregation and weld imperfections, *Welding in the World,* **18,** 89, 1980

5. Séférian, D., *The Metallurgy of Welding,* Chapman and Hall, 1962

6. Gapchenko, M.N., *Automatic Welding,* (*USSR*), **23,** 36, 1970

7. Pearce, B.P. and Kerr, H.W., Grain refinement in magnetically stirred GTA welds of aluminium alloys, *Metall. Transactions,* **12B,** 479, 1981

8. Willingham, D.C. and Bailey, N., *Chemical Grain Refinement of Submerged Arc Welds,* The Welding Institute, Report No M/83/75, July 1975

9. Nilsson, K., Navara, E., Field, J.E. and Easterling, K.E., Avoiding cracks and improving toughness in electroslag welding: a new technique, *Preventing Cracking in Weld Metals, (Conference Proceedings),* Ed. Hrivnák, I., High Tatras, 1981

10. Abson, D.J. and Dolby, R.E., Microstructural transformations in steel weld metal — a reappraisal, *Welding Institute Research Bulletin,* 202, July 1978

11. Åström, H., Loberg. B., Bengtsson, B. and Easterling, K.E., Hot cracking and micro-segregation in 18–10 stainless steel welds, *Metal Science Journal,* 225, July 1976

12. Kyröläinen, A. and Porter, D.A., Precipitation in niobium-stabilized 19 Cr, 11 Ni, 2.6 Mo stainless steel weldments and its effect on impact toughness, *Metal Science Journal,* to be published, 1983

13. Räsänen, E. and Tenkula, J., Phase changes in the welded joints of constructional steels, *Scandinavian Journal of Metals,* **1,** 75, 1972

14. McMahon, J. and Thomas, G., Development of economical, tough, ultra high strength Fe–Cr–C steels, *The Microstructure and Design of Alloys (Conference Proceedings),* vol. 1, p. 180, Institute of Metals, London, 1973

15. Bailey, N., *Effect of Wire Composition and Flux on Solidification Cracking when Submerged-arc Welding C-Mn Steels,* The Welding Institute, Report No 26/1976/M., Abington, 1976

16. Garland, J.G. and Bailey, N., *Solidification Cracking during the Submerged Arc Welding of C-Mn Steels – a Detailed Assessment of the Effect of Parent Plate Compositions,* The Welding Institute, Report No 28/1976/M, Abington, 1976

17. Honeycombe, R.W.K., *Steels — Microstructure and Properties,* Edward Arnold, 1981

18. Dolby, R.E., The influence of niobium on the microstructure and toughness of ferritic weld metal — a review, *Metal Construction,* **13,** 699, 1981

19. Ricks, R.A., Barritte, G.S. and Howell, P.R., Influence of second phase particles on diffusional transformations in steels, *Solid State Phase Transformations (Conference Proceedings),* eds Aaronson, H.I. and Wayman, C.M., Metall. Society of AIME, Pittsburgh, 1981

20. Weatherly, G.C., A determination of the punching stress at the interface of particles during rapid quenching, *Metal Science Journal,* **2,** 237, 1968

21. Billy, J., Johansson, T., Loberg, B. and Easterling, K.E., Stress-relief heat treatment of submerged-arc welded microalloyed steels, *Metals Technology*, 67, Feb, 1980

Further reading

Castro, R.J. and de Cadenet, J.J., *Welding Metallurgy of Stainless and Heat Resisting Steels,* Cambridge University Press, 1968

Christian, J.W., *Theory of Transformations in Metals and Alloys,* Pergamon Press, 1965

Davies, G.J. and Garland, J.G., Solidification structures and properties of fusion welds, *International Metals Review,* **20,** 83, 1975

Dolby, R.E., *Factors Controlling Weld Toughness – The Present Position,* The Welding Institute, Report No 14/1976/M, Abington, 1976

Porter, D.A. and Easterling, K.E., *Phase Transformations in Metals and Alloys,* Van Nostrand Reinhold, 1981

Savage, W.F., Solidification, segregation and weld imperfections, *Welding in the World,* **18,** 89, 1980

Chapter 3

The heat-affected zone

When structural members are joined by fusion welding the material of the plates has to be heated to its melting point and then cooled again rapidly under conditions of restraint imposed by the geometry of the joint. As a result of this very severe thermal cycle the original microstructure and properties of the metal in a region close to the weld are changed. This volume of metal, or zone, is usually referred to as the *heat-affected zone* (HAZ).

The HAZ can be conveniently divided into a number of sub-zones (depending on the material being welded) and this is illustrated for the case of a transformable steel in *Figure 3.1*. Each sub-zone refers to a different type of

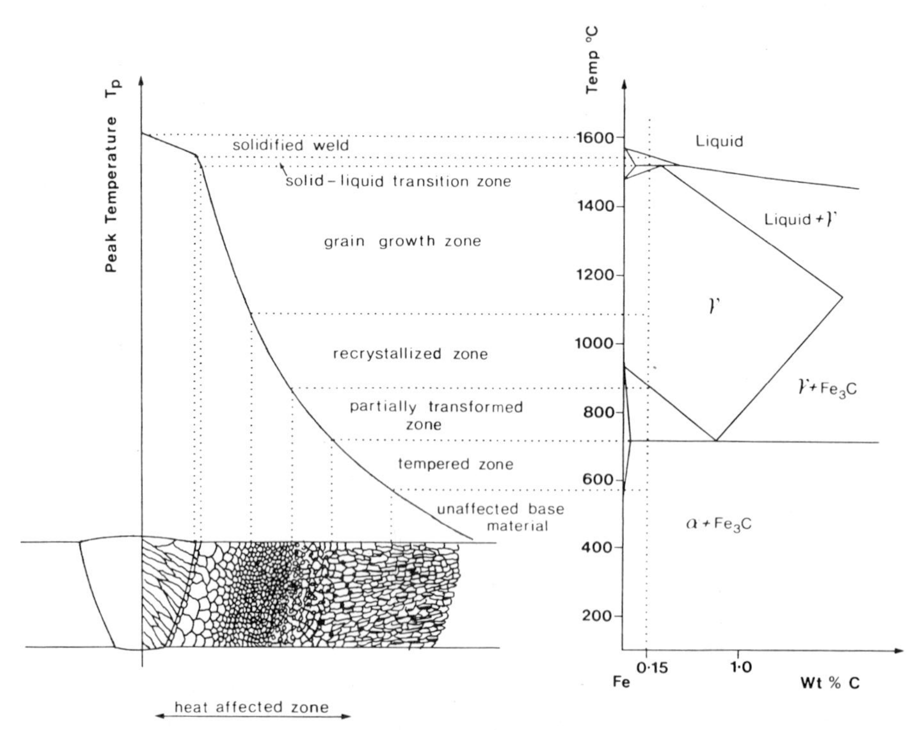

Figure 3.1 A schematic diagram of the various sub-zones of the heat-affected zone approximately corresponding to the alloy C_0 (0.15 wt % C) indicated on the Fe–Fe_3C equilibrium diagram. Compare with *Figure 3.28*

microstructure and, perhaps more important, each structural type is likely to possess different mechanical properties. The structure type and its sub-zone width are partially determined by the thermal cycle, i.e. the complete cycle of heating and cooling due to the movement of the arc and the thermal properties of the base metal. However, the changes in the HAZ are also dependent upon the prior thermal and mechanical history of the material. For example, the recrystallization behaviour during the heating cycle is affected by whether the original material was in cold rolled or annealed condition prior to welding. The onset and extent of the grain growth zone is influenced by the presence of precipitates and their solubility at high temperatures. Thus, in order to obtain a reasonable understanding of the HAZ it is necessary to consider how the microstructure of the base metal reacts to the complete thermal cycle, i.e. the heating cyclc, the time held at temperature (dwell time, including the effect of the melt zone at the fusion line) and, finally, the cooling cycle and its effect on, e.g., phase transformations and precipitate reactions. As before, steel is considered herein because of its technical importance and because of its more complex behaviour due to phase transformations and the presence of carbonitrides. However, the principles established for following structural changes in steels can readily be applied to other materials.

The base material

The bulk of weldable structural steels are normally produced by one of the following manufacturing routes:

1. *Hot rolled,* without subsequent heat treatment.
2. *Control rolled,* i.e. rolled at a temperature sufficiently low that a fine grain size is produced.
3. *Normalized,* following hot rolling by, e.g., forced air cooling.
4. *Normalized,* after cooling to ambient temperature.
5. *Normalized and tempered.*
6. *Quenched and tempered,* to produce a tempered martensitic structure.
7. *Controlled quench,* to produce a bainitic structure.

The range of strengths so obtained lie approximately between 200—600 MPa. For example, in the case of *microalloyed steels,* the strength ranges are usually specified as:

$\sigma_y\,(\text{min}) \approx 360-450$ MPa;
$\sigma_t\,(\text{min}) \approx 550-700$ MPa;
$e\,(\text{min}) \approx 20\%$

where σ_y, σ_t and e refer to yield strength, tensile strength and elongation. In order to meet these specifications, the processing routes have to be based on a rather sophisticated treatment which involves the use of small additions of, e.g. Nb, Ti, V, Al and Zr, or combinations of these, together with C and N. Thus, while the total amount of alloying additions (not counting Mn and Si) rarely exceeds *ca.* 0.15 wt %, the carbonitrides formed are so finely dispersed that relatively thick plates can be delivered in the normalized or controlled rolled condition as *fine grained* steels with good notch toughness. In fact, it is usual to specify a minimum impact resistance of *ca.* 28 J at temperatures down to −40 °C.

To achieve this combination of strength and toughness the grain size has to be uniform and preferably equiaxed. Typical ferrite grain sizes for structural steels are in the range:

1. *ca.* 15—20 μm for plain C–Mn steels.
2. *ca.* 10 μm for C–Mn–Al normalized steels.
3. *ca.* 5 μm for C–Mn–Nb–Al normalized steels.
4. *ca.* 2—3 μm for quenched and tempered steels.

On the basis of the Hall–Petch equation (3.1) σ_y is related to grain size, *d,*

$$\sigma_y = \sigma_i + kd^{-1/2} \tag{3.1}$$

where σ_i refers to the frictional resistance of the lattice to moving dislocations and is a constant for a given material, and k represents the slope of σ_y *vs* $d^{-1/2}$. However, in the case of many microalloyed and controlled rolled steels, the grain size only accounts for about half the yield strength. For example, contributions to strength have been suggested[1] for a Nb-microalloyed controlled

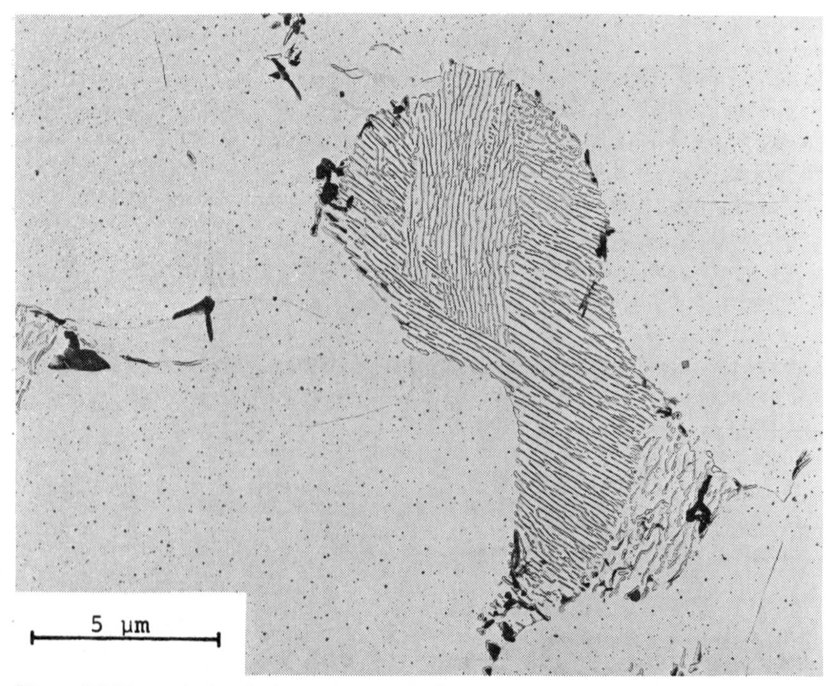

Figure 3.2 Transmission electron micrograph of a carbon extraction replica of carbides and a pearlite colony in a Ti–V microalloyed normalized steel. Courtesy of A. Nordgren, University of Luleå

rolled steel (these can, in fact, vary considerably from steel to steel) as shown in equation (3.2), where σ_{text} refers to the contribution due to texture.

$$\sigma_y\,(550\text{ MPa}) = \sigma_i\,(6\%) + \sigma_{solid\ soln}\,(25\%) + \sigma_{ppt\ hard}\,(6\%) + \sigma_{text}\,(8\%) + \sigma_{disl}\,(8\%) + kd^{-1/2}\,(47\%) \quad (3.2)$$

Note that, in spite of the presence of fine dispersions of Nb(CN), the contribution from precipitation hardening is only *ca.* 35 N mm^{-2}. Gladman, Dulieu and McIver[2] have applied the Ashby–Orowan model of precipitation hardening to steel to obtain expression (3.3), where V_f is the volume fraction

$$\sigma_{ppt\ hard} = \frac{5.9\,(V_f)^{1/2}}{\bar{x}} \times \ln\left(\frac{\bar{x}}{2.5 \times 10^{-4}}\right) \quad \text{MPa} \quad (3.3)$$

and $\bar{x}$ the mean-planar-intercept path of (or mean distance between) precipitates. Measurements from transmission electron micrographs of

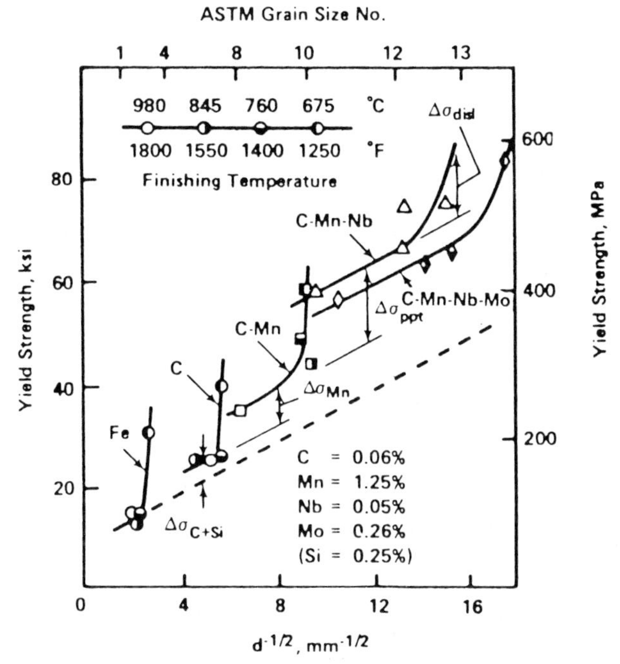

Figure 3.3 Relationship between yield strength and grain size for a number of steels. After Porter, L.F. and Repas., P.E., The evolution of HSLA steels, *Journal of Metals,* 14, April 1982

extraction replicas of mixed TiN and VN precipitates in a microalloyed normalized steel (*Figure 3.2*) gave an $\bar{x}$ of *ca.* 100—200 nm for particle sizes of *ca.* 10—30 nm, and $V_f \approx 0.004$–5 (very approximate), which, according to equation (3.3) gives a $\sigma_{ppt\ hard}$ of *ca.* 10—15 MPa. This low value is in reasonable agreement with equation (3.2) and shows that the role of carbides in many steels is to 'pin' grain boundaries, i.e. to help refine the grain size. There are, however, exceptions; for example VC gives substantial precipitation hardening in certain steels. As a general rule NbC, VN and TiN are mainly present to give grain refinement.

Dependence of the strength of steels on grain size is shown in *Figure 3.3* for a number of different steels. *Figure 3.4* illustrates that a similar proportionality exists between grain size and fracture stress, σ_f. This relationship has been established by Cottrell[3] and is given by equation (3.4) where σ_i' is the modified

$$\sigma_f = \tfrac{1}{2}\,\sigma_i' + 2\,(\mu\gamma_s)\,d^{-1/2} \tag{3.4}$$

frictional stress, μ is the shear modulus and γ_s is the surface energy of the metal. While equation (3.4) essentially gives the fracture stress for a ductile fracture, it is useful to understand the effect of grain size and frictional stress on the *transition temperature* of steel, i.e. the critical temperature that marks the

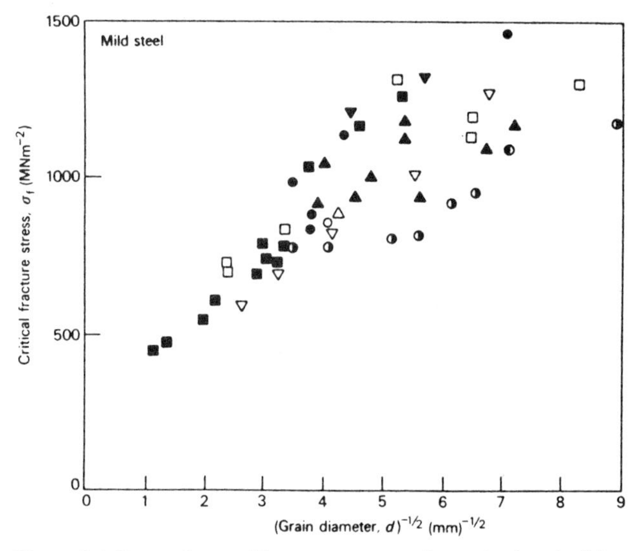

Figure 3.4 Dependence of fracture stress on the grain size of mild steel. Taken from Honeycombe, R.W.K., *Steels–Microstructure and Properties*, Edward Arnold, 1981 (quoting work of J.F. Knott)

transition between ductile and brittle cleavage behaviour. In the case of ductile materials, $\sigma_y > \sigma_f$, but as the temperature is lowered, σ_y increases at a faster rate than σ_f. Thus, the transition temperature corresponds to the cross-over point at which $\sigma_y = \sigma_f$ and this is illustrated in *Figure 3.5* (*a*). The effect of grain refinement, on the other hand, tends to raise σ_f more than σ_y, which therefore lowers the transition temperature, as indicated in *Figure 3.5* (*a*); this again emphasizes the significance of grain refinement in structural steels. The effect of grain refinement on transition temperature in a number of steels is shown in *Figure 3.5* (*b*).

Another factor which is of importance in the fracture toughness of metals is ductility. In fracture toughness this is usually expressed as a function of some critical crack length to mark the onset of brittle failure. Using a modified Griffith equation, it can be shown that the condition for a crack to grow in a

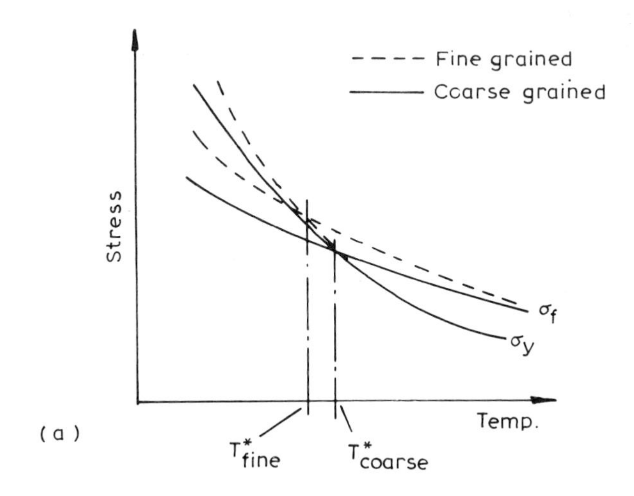

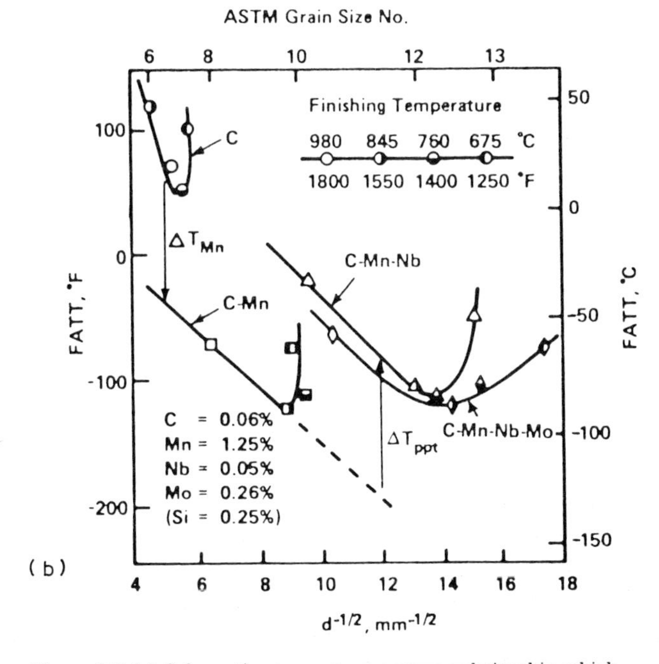

Figure 3.5 (a) Schematic stress–temperature relationship which compares fine and coarse grained metals. The cross-over between the yield and cleavage fracture stresses defines the ductile–brittle transition temperature. (b) Relationship between grain size and transition temperature in a number of steels. After Porter, L.F. and Repas, P.E., The evolution of HSLA steels, *Journal of Metals,* 14, April 1982

metal is given by equation (3.5), where c^* is the critical crack radius and γ_p

$$c^* \gtrsim \frac{E\,(\gamma_p + 2\gamma)}{\pi \sigma_f^2} \qquad (3.5)$$

refers to the *plastic work of fracture,* i.e. the additional work needed to reopen a blunted crack; γ refers to the metal's surface energy. This equation shows that raising σ_f, e.g. by hardening the matrix and increasing σ_i', effectively decreases the critical crack length for brittle fracture. Fortunately, structural steels are normally ductile which means that γ_p is very large and c^* is correspondingly increased.

It is interesting to rearrange equation (3.5) to give equation (3.5a). This

$$\sigma_f = \left[\frac{E\,(\gamma_p + 2\gamma)}{\pi c}\right]^{1/2} \qquad (3.5a)$$

form of the equation implies that as a crack starts to grow c increases, thereby lowering σ_f. However, since the applied stress remains constant, crack growth will accelerate. In fact, it can be shown that the crack velocity is given by equation (3.6)[4], where c is the crack size at a given instant, ϱ is the material

$$v = k\left[\frac{E}{\varrho}\left(1 - \frac{c^*}{c}\right)\right]^{1/2} \qquad (3.6)$$

density and k is a constant. Thus, the velocity increases rapidly as c increases and may even attain about half the speed of sound in a solid, i.e. *ca.* 1000 m s^{-1}! Such rapid fracture phenomena have, in fact, occurred in recent years in pipe lines[5] in which cracks were estimated to propagate at speeds of *ca.* 500—600 m s^{-1} for distances of *ca* 600 m. Even ductile fractures have been reported to reach speeds of *ca.* 60 m s^{-1}.

Note that in most metals, d in equation (3.1) refers to the mean grain size. In bainitic and quenched and tempered steels d refers to the mean *width* of the bainite or martensite laths. In the case when equation (3.4) is used to describe the fracture stress of the steel, however, d cannot be expressed in terms of the lath widths. The reason for this is that adjacent laths are separated only by small angle boundaries, i.e. the crystallographic orientation between adjacent laths differs by only a few degrees. While such boundaries provide a fairly effective barrier to dislocation pile-ups, they are not able to arrest a propagating crack. Thus, lath boundaries can contribute to strength, but only a little to fracture toughness. In computing the σ_f of bainite or quenched and tempered steels, therefore, the lower limit is given by the prior austenitic grain size and the upper limit by the distance between lath packets.

The base metal's carbon equivalent

An essential requirement of any structural material to be welded is that it has good weldability. The weldability of steels is usually expressed in terms of a *carbon-equivalent* (C_{equiv}) limit, or maximum value. As a general rule, a steel is

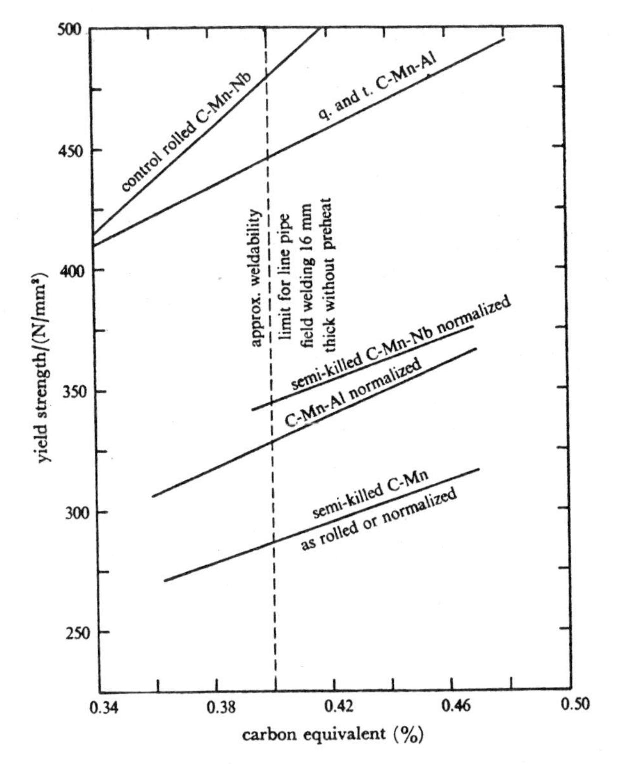

Figure 3.6 Plots of lower yield strength *vs.* carbon equivalent composition for a number of structural steels. The vertical dotted line corresponds to $C_{equiv} = 0.4$ and defines the approximate upper limit of hardness if the steel is to possess good weldability. After Baker, R.G., Weldability and its implications for material requirements, *Rosenhain Centenary Conference Proceedings*, p. 129, eds Baker, R.G. and Kelly, A., Royal Society, 1975

considered[6] weldable if $C_{equiv} < 0.4$. Its value is assessed in terms of how the alloying elements present affect the transformation characteristics, including the M_s temperature, of the steel, i.e. whether C-curves of the CCT diagram move to longer or shorter times. In effect, the C_{equiv} provides an indication of the type of microstructure to be expected in the weld heat-affected zone as a function of the cooling rate from peak temperature. More specifically, it is supposed to give an indication of whether or not martensite is able to form, i.e. of the *hardness* of the weld. The carbon-equivalent formula (3.7) has been

$$C_{equiv} = C + \frac{Mn}{6} + \frac{Cr + Mo + V}{5} + \frac{Cu + Ni}{15} \tag{3.7}$$

adopted by sub-commission IX-G of the International Institute of Welding (all amounts in wt %). An alternative formula, equation (3.8), has been adopted in

$$C_{equiv} = C + \frac{Si}{24} + \frac{Mn}{6} + \frac{Ni}{40} + \frac{Cr}{5} + \frac{Mo}{4} + \frac{V}{14} \tag{3.8}$$

Japan. Several other variants are available and depend on the alloying content of the steel (*see* Appendix, *Table A2*). Since high hardness and the presence of martensite are known to influence what is considered the most serious type of weld defect, *cold cracking* or *hydrogen cracking,* a C_{equiv} of 0.4 is claimed to be the maximum allowable value if these defects are to be avoided. *Figure 3.6* shows σ_y as a function of C_{equiv} for a number of different steels. Thus, the value of 0.4 is observed to set the upper limit of good weldability without the need for preheat. *Figure 3.6* shows that good weldability is not necessarily a function of the strength of the steel, but more an indication of how alloying affects the hardenability of the metal during the cooling cycle of the weld. It should also be noted that measurements which relate strength or hardness to C_{equiv} exhibit considerable scatter, and this implies that caution needs to be used when applying these concepts in practice.

Clearly, the use of empirical formulae such as equations (3.7) and (3.8) to describe weldability is not very satisfactory. As pointed out by Cotton[7], for example, it seems timely to question the accuracy of C_{equiv} formulae when applied to low carbon structural steels of the type used, e.g., in pipelines. The practice can even be extremely expensive when preheat is required, irrespective of other considerations such as plate thickness, welding conditions, hydrogen level of electrodes, geometry of joint, etc., all of which can affect the microstructure and susceptibility to cold cracking. In particular, the effect of the elements Mn and Mo are likely to be overestimated in these reduced C-steels with regard to their influence on hardening. The C–Mn and C–Mn microalloyed steels are typically based on alloying additions of:

0.22 wt % C (max); *ca.* 0.30 wt % Si; *ca.* 0.5 — 1.5 wt % Mn;
0.04 wt % P (max); 0.02 wt % S (max)

Using the maximum levels quoted, the C_{equiv} (from equations (3.7) and (3.8)) has a value of *ca.* 0.48. Indeed, the steel manufacturers have the problem of trying to produce higher strength steels while strictly limiting the C and Mn levels in order to keep the C_{equiv} below 0.4.

The heating cycle

In many respects the heating cycle is at least as important as the dwell time and cooling cycle in terms of influence on the final microstructure and properties of the HAZ. In spite of this, as discussed in the section on weld simulation (pp. 29—33), the heating cycle is often considered of secondary importance when the welding simulator is programmed. Most attention is usually given to obtaining a correct Δt_{8-5}. In steels, for example, the rate of heating to the peak temperature determines the following important parameters:

1. The temperature of recrystallization.
2. The degree of superheating, $-\Delta T$, in the $\alpha \rightarrow \gamma$ phase transformation.

3. The rate of coarsening of carbides and nitrides.
4. The temperature of solution of carbides and nitrides.
5. The *main* proportion of grain growth.

The importance of these factors is in their ability to affect the degree of grain growth in the HAZ, since it is this more than anything else which affects the final properties of the weld. Some of these parameters are considered in more detail below.

Recrystallization

Because of the $\alpha \rightarrow \gamma$ phase transformation which occurs on heating transformable steels, recrystallization is probably of most interest in the present context in connection with certain austenitic stainless steels and other metals that do not undergo a phase transformation. Recrystallization is dependent mainly upon three variables:

1. The temperature at which it occurs.
2. The amount of prior deformation.
3. The purity of the metal.

The temperature of recrystallization determines the rate of nucleation and of growth of new grains, equation (3.9), where Q_r is the activation energy for

$$\text{Rate} = A \exp(-Q_r/RT) \tag{3.9}$$

recrystallization and R is the gas constant. In practice, the actual temperature at which recrystallization occurs is very much a function of prior deformation of the material. Indeed, some metals may not recrystallize at any temperature up

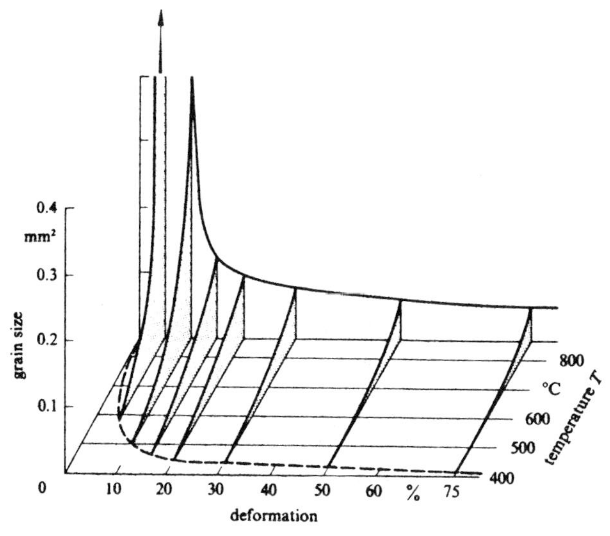

Figure 3.7 The inter-dependence of recrystallization temperature, prior deformation and initial grain size. After Haasen, P., *Physical Metallurgy*, Cambridge University Press, 1978

to T_m without prior deformation since a vital prerequisite for recrystallization i that polygonization occurs, i.e. a rearrangement of dislocations to form nev boundaries. These boundaries then move under the driving force of the storec energy due to deformation. As a rough approximation, the energy stored, E, i proportional to the dislocation density, ϱ, equation (3.10). However, it i

$$E \propto \varrho \tag{3.10}$$

shown by experiment that the recrystallized grain size is fairly independent o the temperature, provided some minimum deformation has already occurred[8] This critical amount of deformation need only be a few percent in fine graine metal. The recrystallization temperature is strongly affected by the impuritie present since these can effectively hinder polygonization. An example of recrystallization diagram for pure iron, which illustrates the relative depen dence between prior deformation and temperature, is shown in *Figure 3.7* Clearly, relatively small deformations and high temperatures give the mos grain growth. On the other hand undeformed metals do not appear to recrystal lize at all. Indeed, the occurrence of recrystallization and grain growth in th HAZ of 18/8 type stainless steel welds is known to be dependent on the plate being in a cold-worked condition prior to welding[9].

The α→γ phase transformation

The heating rate in many welding processes can be very high. Fo example, in normal arc welding it can be of the order of *ca.* 200—300 °C s^{-1}

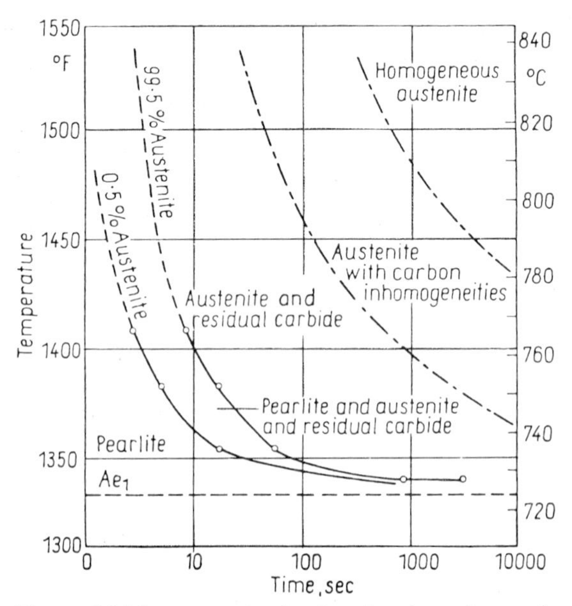

Figure 3.8 Measurements showing the dependence of austenitizing on temperature and time. After Roberts, G.A. and Mehl, R.F., *Transactions AIME*, **154**, 318, 1943

The implication is that the $\alpha \rightarrow \gamma$ phase transformation not only occurs above T_0 (the equilibrium temperature as given by the Fe–Fe_3C phase diagram), but also that the α-phase is likely to be substantially superheated before the transformation occurs. Furthermore, the degree of superheating is affected by the welding process, i.e. the energy input of the weld. Some measurements that illustrate this effect are shown in *Figure 3.8*, where it is shown that considerable superheating occurs at the fastest heating rates, i.e. short heating times. *Figure 3.8* also gives an estimate of how the presence of carbides may affect the extent of the $\alpha \rightarrow \gamma$ transformation. Another surprising result shown in *Figure 3.8* is that the dissolution of carbides (presumably cementite in this case) occurs at temperatures much greater than their expected solubility temperature, particularly with short heating times. This is an important point which is returned to shortly.

Consideration of the rapid heating rates in welding has lead to the suggestion[10] that since the $\alpha \rightarrow \gamma$ transition is, in some cases, likely to occur by a reverse martensitic transformation, the resulting deformation due to the transformation strains may have the effect of accelerating primary recrystallization and grain growth in the HAZ of steels.

Precipitate stability

The coarsening and dissolution of precipitates is a diffusion-controlled process in which time and temperature are mutually dependent. In the case of carbides or nitrides in steels the process is usually complicated by the presence of substitutional elements other than iron, and in this case it seems likely that the diffusivity of the carbide former at the temperature considered is rate-controlling.

The equilibrium solubility of carbides and nitrides in steel and how the particle solubility temperatures may change during a rapid heating cycle are

TABLE 3.1 Solubility products* for carbides and nitrides in steels (after Ashby and Easterling[11])

Compound	*Metal*	*Non-metal*	*A*	*B*
$Cr_{23}C_6$	Cr	C	5.90	7375
V_4C_3	V	$C_{0.75}$	5.36	8000
TiC	Ti	C	2.75	7000
NbC	Nb	C	2.96	7510
Mo_2C	Mo	C	5.0	7375
Nb(CN)	Nb	(CN)	2.26	6770
VN	V	N	2.27	7070
VN	V	N	3.40 +0.12 Mn	8330
AlN	Al	N	1.80	7750
AlN	Al	N	1.03	6770
NbN	Nb	N	4.04	10230
NbN	Nb	N	3.70	10800
TiN	Ti	N	0.32	8000

*The data refer to dissolution of M_aC_b or M_aN_b, following the equation:

$$\log_{10} [C_M]^a [C_C]^b = A - B/T$$

with concentrations measured in wt % and T in K

considered. Data exist for the equilibrium solubility products, (C_MC_C) or (C_MC_N) in austenite, for most important carbides or nitrides in steels. Here C_M is the concentration of the *carbide or nitride former* (e.g. Nb), while C_C and C_N represent the concentrations of carbon and nitrogen (M is the metal). The equilibrium solubility product for the carbide-producing reaction (3.11)

$$a\mathrm{M} + b\mathrm{C} = \mathrm{M}_a\mathrm{C}_b \tag{3.11}$$

(where a and b effectively define the stoichiometry of the carbonitride) is given by equation (3.12), where A and B are constants derived from *isothermal*

$$\log\,[C_\mathrm{M}]^a[C_\mathrm{C}]^b = A - \frac{B}{T} \tag{3.12}$$

experiments. The product $[C_\mathrm{M}]^a[C_\mathrm{C}]^b$ may be expressed in wt % or atomic %. Values of A and B for a number of carbides and nitrides in austenite are given in *Table 3.1.* Some of these are presented graphically in *Figure 3.9,* which shows that the particles progressively dissolve with increase in temperature. Thus, by

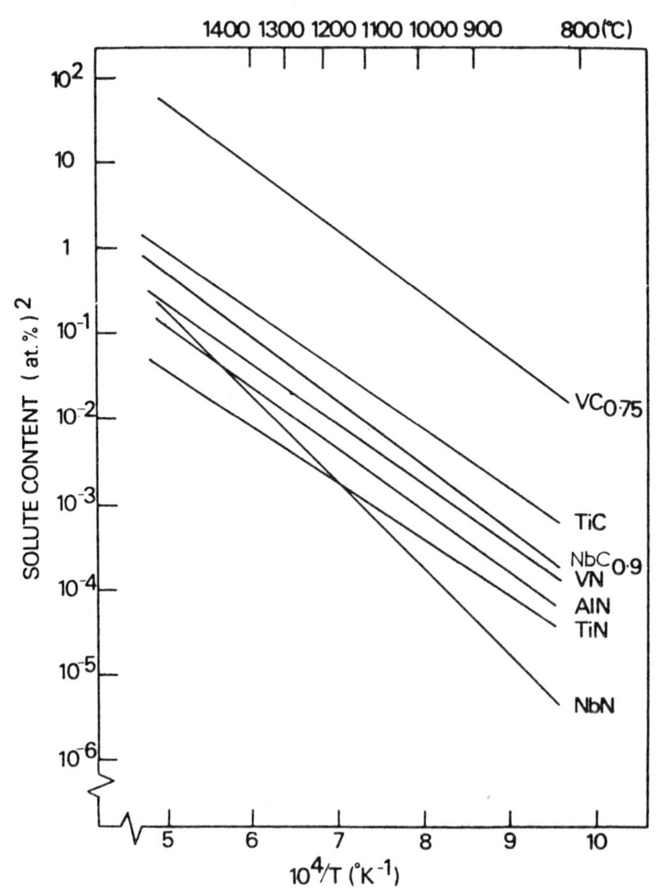

Figure 3.9 Solubility products of carbides and nitrides in austenite as a function of temperature. After Aaronson, B., *Steel Strengthening Mechanisms,* Climax Molybdenum Co., 1969

knowing the total percentages of the combining elements in the steel involved in the reaction, the temperature of complete solution of the particles can be obtained from *Figure 3.9*.

The lower solubility of, e.g., TiN particles (as shown by *Figure 3.9*) compared, for example, with AlN or VN, is because of their *free energy of formation*, ΔH_f. The free energy of formation for various carbides, nitrides and borides is shown in *Figure 3.10*. In general, the more negative is ΔH_f, the less soluble the carbide or nitride tends to be, although there are exceptions to this rule. It is interesting to note that some of the borides appear to exhibit about the same stability as the most stable nitrides.

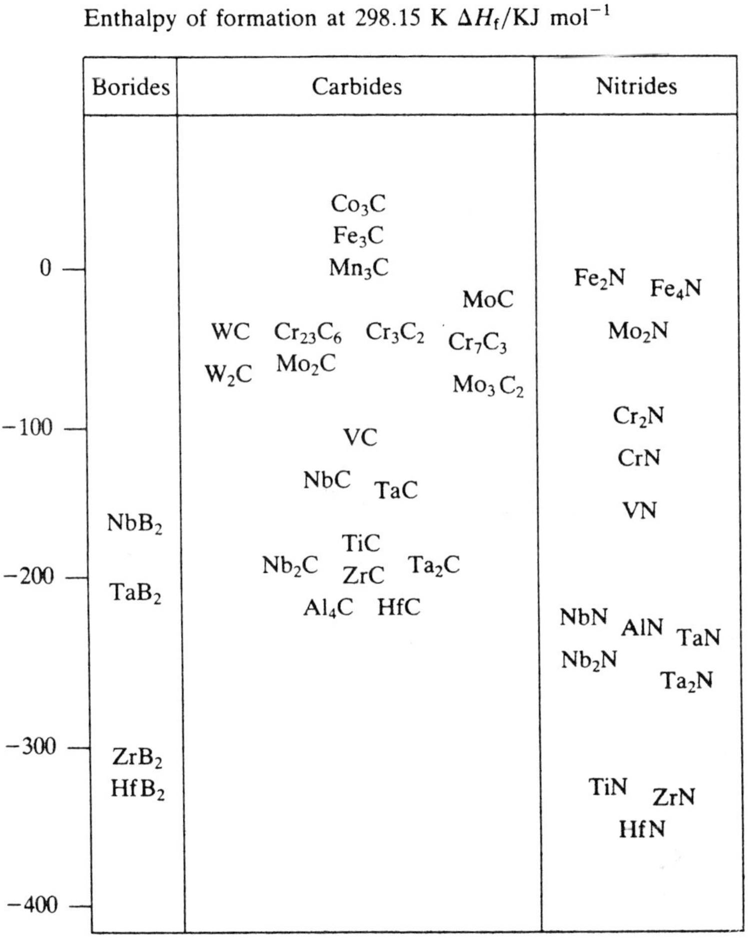

Enthalpy of formation at 298.15 K ΔH_f/KJ mol^{-1}

Borides | Carbides | Nitrides

Co_3C Fe_3C Mn_3C
0
MoC Fe_2N Fe_4N
WC $Cr_{23}C_6$ Cr_3C_2 Cr_7C_3 Mo_2N
W_2C Mo_2C
Mo_3C_2
-100 Cr_2N
VC
CrN
NbC TaC
VN
NbB_2
TiC
-200 Nb_2C ZrC Ta_2C
TaB_2
Al_4C HfC
NbN AlN TaN
Nb_2N
Ta_2N
-300 ZrB_2
HfB_2 TiN ZrN
HfN
-400

Figure 3.10 Enthalpies of formation of carbides, nitrides and borides. After Schick, H.L., *Thermodynamics of Certain Refractory Compounds,* Academic Press, 1966

The solubility limits calculated from the data in *Table 3.1* are not generally correct for welding except, perhaps, at the highest energy inputs. In most cases equilibrium is not reached at the heating rate used and some superheating occurs. It is important to estimate what this may be in view of the significance of 'particle pinning' in grain growth. In doing so two possibilities should be considered:

1. That the particles do not dissolve completely during the weld cycle, but that some particle coarsening occurs.
2. That the particles do dissolve completely during the heating cycle.

The former case is of particular relevance to recently developed steels that contain fine dispersions of TiN particles. The latter is of relevance to most other steels used in practice.

Precipitate coarsening during a weld thermal cycle

Examples of TiN precipitates in a Ti-steel after high energy input weld simulation are shown in *Figure 3.11*. Changes in particle size distribution before and after weld simulation are also included in the figure, which illustrates that the average particle size following weld simulation increases from *ca.* 8 nm to *ca.* 14 nm. Experiments have shown that the coarsening rate is usually accelerated if the particles have mixed composition, e.g. Ti(V)(N).

A well known relationship has been derived, based on the Lifshitz–Wagner theory for Ostwald ripening[12], in which the change in particle size after a weld thermal cycle $T(t)$ can be modified to equation (3.13), where r_o and r are

$$r^3 - r_o^3 = \int_0^\infty \frac{A}{T(t)} \exp - \left(\frac{Q_V}{RT(t)}\right) dt = I \tag{3.13}$$

the original and new particle radii, A is a constant depending on the matrix composition and Q_v is the activation energy for diffusion between particles.

Equation (3.13) assumes that volume diffusion is rate-controlling. When applying equation (3.13) to a weld thermal cycle, the precipitate coarsening is given as a function of both temperature and time. A convenient method of estimating the amount of coarsening after a weld thermal cycle is to measure the amount of coarsening due to a known thermal cycle and compare the integrated areas under the appropriate temperature–time cycles. Thus, knowing the mean particle size, r^*, and the thermal cycle, $T^*(t)$, equation (3.13) may be written as equation (3.14). The unknown kinetic constants in A

$$(r^*)^3 - r_o^3 = \int_0^\infty \frac{A}{T^*(t)} \exp - \left(\frac{Q_V}{RT^*(t)}\right) dt^* = I^* \tag{3.14}$$

are then eliminated by dividing equation (3.13) by equation (3.14) to give equation (3.15). In this way, the required values of r can be obtained for any

$$\frac{r^3 - r_o{}^3}{(r^*)^3 - r_o{}^3} = \frac{I}{I^*} \tag{3.15}$$

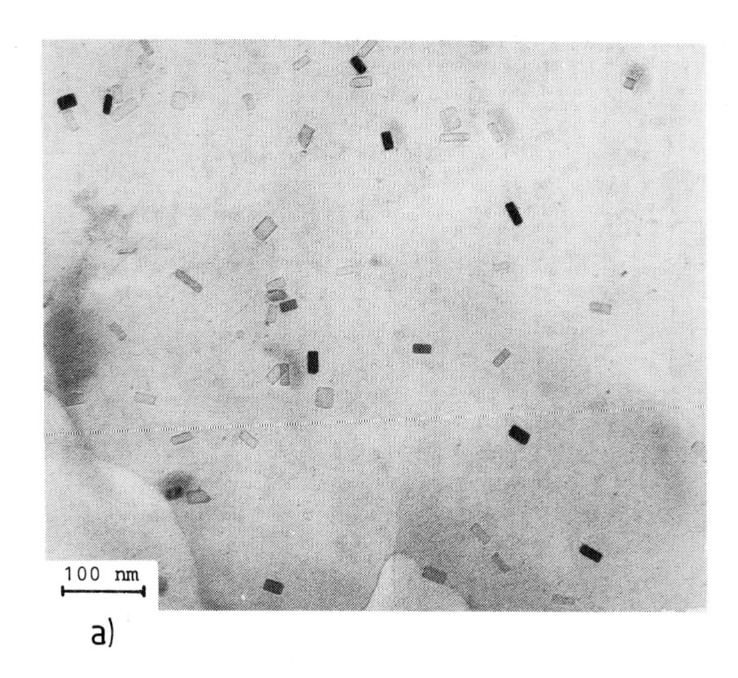

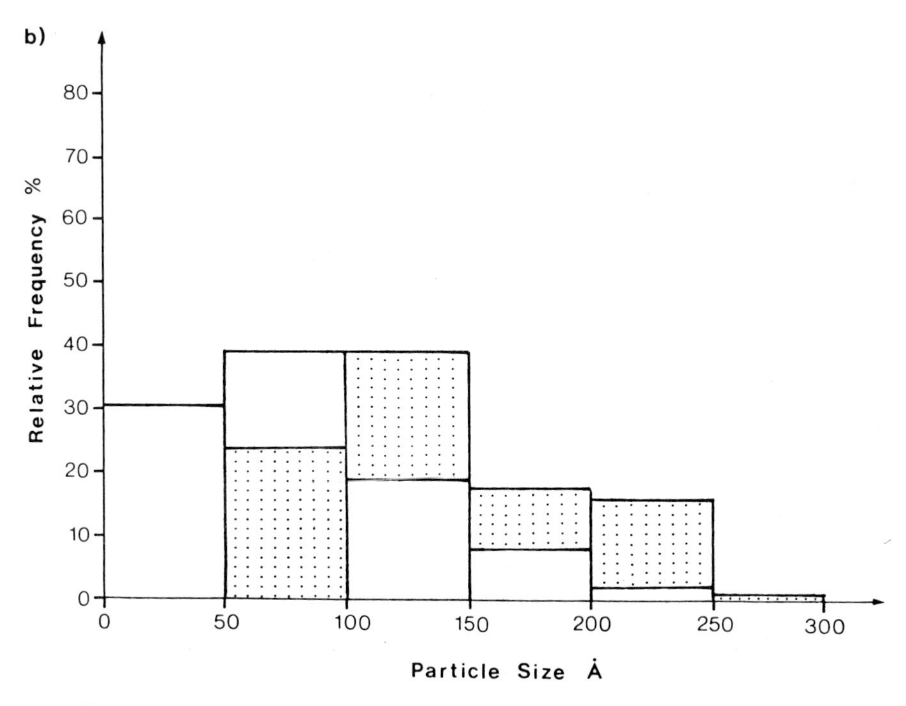

Figure 3.11 (a) Carbon-extraction replica and (b) measurements (from carbon-extraction replicas) of changes in size distributions of TiN precipitates in a normalized Ti-microalloyed steel. The shaded area refers to particle sizes after weld simulation. Courtesy of J. Strid and J. Ion, University of Luleå

given values of T_p and Δt using equations (1.20) or (1.24), provided Q_v is known. Of course, it should be emphasized that this approach is essentially only an approximation and neglects, e.g., changes in solid solubility with increasing temperature.

Applying this approach to the welding of Ti-microalloyed steels, in which coarsening of TiN precipitates occurs, r^* and r_o can now be estimated from *Figure 3.11* in which:

$$2\ r_o \approx 8.2\ \text{nm};\ 2\ r^* \approx 14.1\ \text{nm}$$

and

$$\Delta t^* = 40\ \text{s};\ T_p^* = 1564\ \text{K};\ Q_v/RT_m = 14.5$$

The Q_v/RT_m value is a constant for the diffusion of Ti in austenite (*see Figure 3.13*). Using a computer, contours of particle size as a function of Δt (or welding energy input), and temperature can now be produced, as shown in *Figure 3.12*. The approach used here is similar to that developed for the grain growth diagrams[11] (*see* pp. 126–130). In *Figure 3.12*, the particle size contours

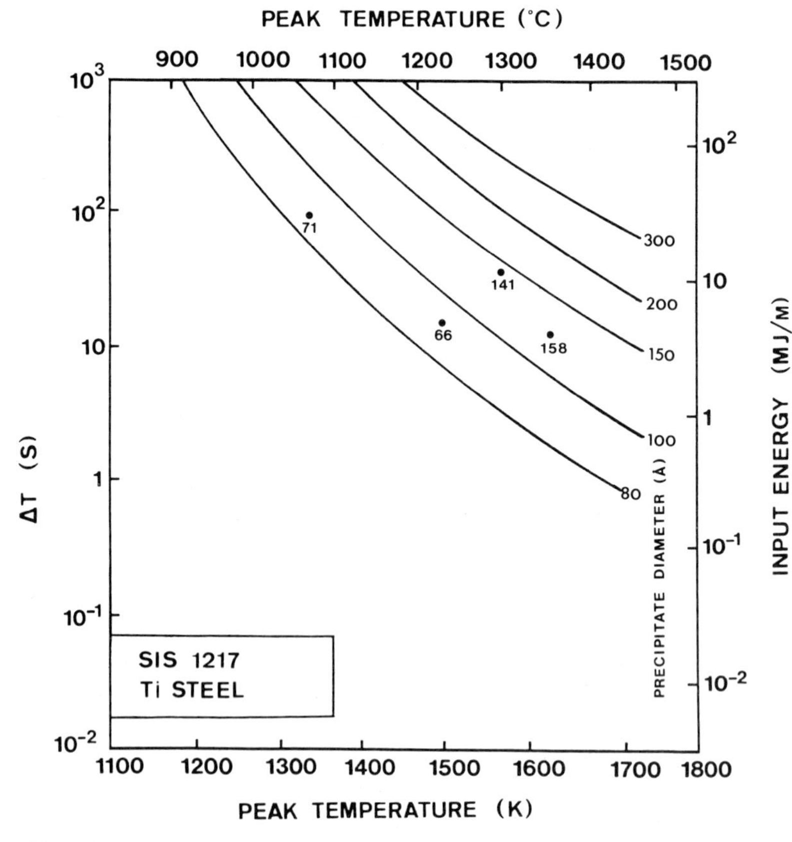

Figure 3.12 Theoretical plot of particle coarsening as a function of welding input energy and peak temperature in the HAZ of a Ti-microalloyed steel. The precipitates in this case are TiN. The points given in the diagram refer to experimental measurements. After unpublished research of J. Strid and J. Ion, University of Luleå

are all drawn relative to $2r^* = 14.1$ nm. However, the experimental points other than 14.1 nm were determined independently of the predicted particle sizes. The fit between theoretical prediction and experiment could be better, but it does at least allow some estimation of how different weld thermal cycles may affect particle coarsening – and hence grain growth in the HAZ of these steels. This is discussed again on pp. 130–132.

Precipitate dissolution during a weld thermal cycle

The theory of precipitate dissolution during a weld thermal cycle[11] is based on the assumption that during a rapid heating cycle the equilibrium solubility of carbonitrides as given by equation (3.12) has to be modified by a suitable factor that accounts for the superheating of the particles. Consider a particle which is unstable and begins to dissolve. It has been well established that a spherical diffusion field of radius l, surrounding a spherical particle, is related to the diffusivity of the particle's metal component in the matrix, D, and the time, t, by equation (3.16). If the particle is to dissolve completely in the time t, the

$$l = (Dt)^{1/2} \tag{3.16}$$

metal and the carbon and nitrogen it contains must now dissolve in the volume $4/3\pi(Dt)^{3/2}$, and this requires some superheating. Thus, due to superheating, only a fraction, f, of the diffusive field is utilized since the particle has not had time to dissolve completely. It can hence be shown that the matrix volume fraction, f, is given by equation (3.17). Note that f has the property that when

$$f = \frac{(Dt)^{3/2}}{l^3 + (Dt)^{3/2}} \tag{3.17}$$

$(Dt)^{1/2}$ is small, $f = (\mathrm{Dt})^{3/2}/l^3$; when it is large, $f \approx 1$. Therefore, the faster the heating rate, the smaller f becomes.

Now, for a given heating rate, the particle dissolves completely when equation (3.18) is satisfied, where D^* and t^* are the diffusivity coefficient and

$$l = (D^* t^*)^{1/2} \tag{3.18}$$

time relevant to the temperature, T^*, at which complete solution occurs. Equation (3.19) follows, where Q_v is the activation energy for diffusion of the

$$f = \frac{1}{1 + \left[\frac{t^*}{t} \exp - \frac{Q_v}{R}\left(\frac{1}{T^*} - \frac{1}{T}\right)\right]^{3/2}} \tag{3.19}$$

metal atom and R is the gas constant. If the temperature at which full solution takes place is defined as T_s then, utilizing equation (3.12), equation (3.20) can

$$T_s = \frac{B}{A - \log\left(\frac{[C_M]^a[C_C]^b}{f^{a+b}}\right)} \tag{3.20}$$

be derived[11], where data for A and B are given in *Table 3.1* (p. 115). To a first approximation a value of Q_v appropriate to the diffusion of the metal atoms in austenite can be assumed. There have been very few measurements of the time taken for a particle to dissolve, t^*, at a given temperature. However, Ikawa *et al.* [13] have estimated a t^* of *ca.* 20 s for NbC dissolution at the equilibrium

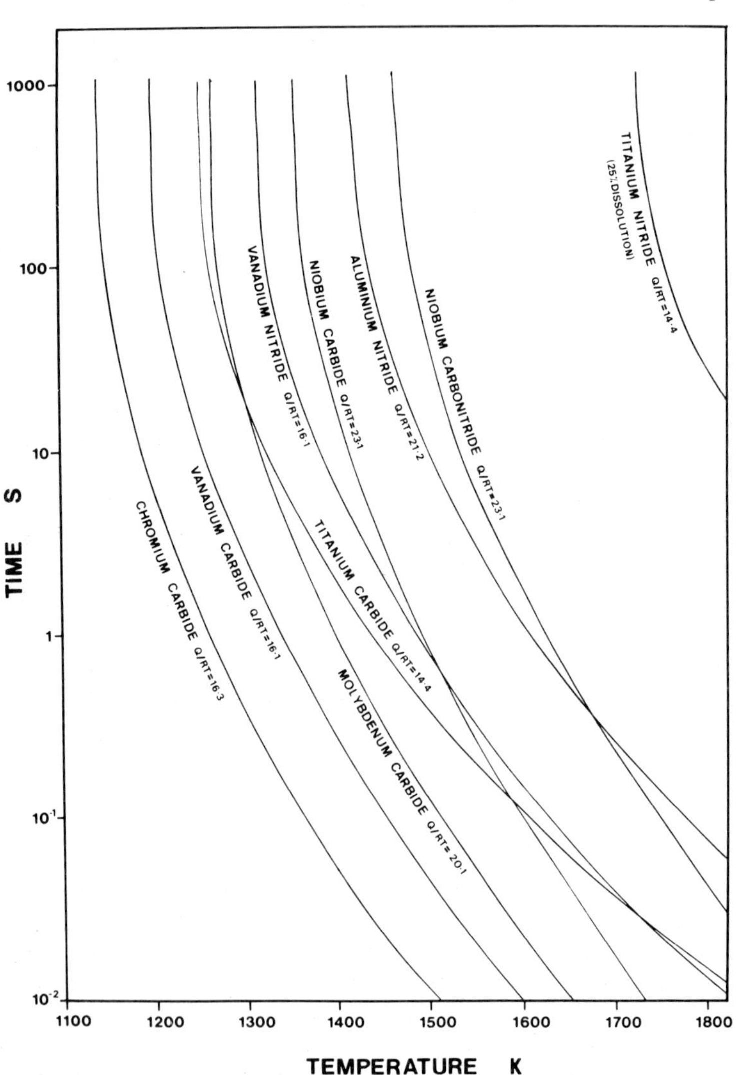

Figure 3.13 The times of complete dissolution of various carbides and nitrides in austenite as a function of temperature, based on the model of ref. 11. The shorter the weld thermal cycle, the greater the superheating of the carbonitrides. The diffusion data are obtained from Mrowec, S., *Defects and Diffusion in Solids — An Introduction,* Elsevier, 1980

temperature of dissolution, and this can be used in equation (3.20). On this basis, the solubility curves for a number of carbides can be plotted as a function of heating time and temperature, as shown in *Figure 3.13*. Values of Q_v/RT_m used for the various particles are given in the figure. The Δt_{8-5} times can be related to actual rates of heating by the Rosenthal equations relevant to bead-on-plate welds of thick plates (equation (1.17)). The degree of superheating increases with heating rate and this has to be taken into account when trying to estimate grain growth in the HAZ of steel welds. The compositions used are all typical of commercial steels. It is interesting to note that the mixed Nb(CN) has a much lower solubility than, e.g., NbC. However, easily the highest dissolution temperature is exhibited by TiN. Indeed, TiN is predicted not to dissolve completely at temperatures up to T_m, in agreement with experimental observation.

So far particle coarsening and dissolution have been considered as separate processes. In practice, of course, coarsening always precedes dissolution, and this may tend to move the curves of *Figure 3.13* to slightly higher temperatures. This is discussed later, *see* pp. 130–132.

Grain growth

It is apparent from the discussion on base material (pp. 105–112) that grain size is of paramount importance in metals, it being a key factor in determining the strength and toughness of a material. It also has enormous importance with respect to determining an alloy's susceptibility to cold cracking and reheat cracking in welds, *see* Chapter 4. The weld thermal cycle is such that in the majority of metals some grain growth occurs in the HAZ and this not only affects strength and toughness, but also influences the grain size of the weld metal (*see* Chapter 2). As a result of its importance, there have been a large number of papers, both experimental and theoretical, that deal with grain growth in the HAZ, although the majority of these have tended to discuss a given alloy rather than to give an overview of the phenomenon. The kinetics of grain growth are considered first and then the relation of grain growth to the weld thermal cycle is discussed.

Kinetics of grain growth

Following primary recrystallization, or completion of the $\alpha \rightarrow \gamma$ phase transformation in steels, the stability of the new grains is far from equilibrium. The fine size of the grains and the fact that it is impossible to achieve mechanical equilibrium between grain boundary tensions in a three-dimensional arrangement of grains mean that the structure contains a certain residual stored energy. Thus, grains with more sides begin to swallow grains with less sides, as illustrated in *Figure 3.14,* and the energy progressively decreases.

If it is assumed that the mean radius of curvature of the grain boundaries (gb) is proportional to the mean grain diameter, $\bar{d}$, the average driving force for grain growth is given by equation (3.21). The mean velocity of growth, $\bar{v}$, is

$$\Delta G \approx \frac{2\gamma_{gb}}{\bar{d}} \text{ per unit volume} \tag{3.21}$$

given by equation (3.22) where M is the *mobility* of the boundary and is highly

$$\bar{v} = \frac{d\bar{d}}{dt} = M \frac{2\gamma_{gb}}{\bar{d}} \tag{3.22}$$

dependent upon the temperature. Integration gives equation (3.23) where

$$\bar{d}^2 = d_o^2 + Kt \tag{3.23}$$

$K \approx M\gamma_{gb}$ and d_o is the original grain size. Experimentally, realtion (3.24) is

$$\bar{d} = K't^n \tag{3.24}$$

obtained, where K' is a proportionality constant. If $n=0.5$ and $d \geq d_o$, then equations (3.24) and (3.23) are equivalent. In practice, however, n only approaches 0.5 in very pure metals, and in alloys it is usually much less than 0.5 due to impurity drag effects.

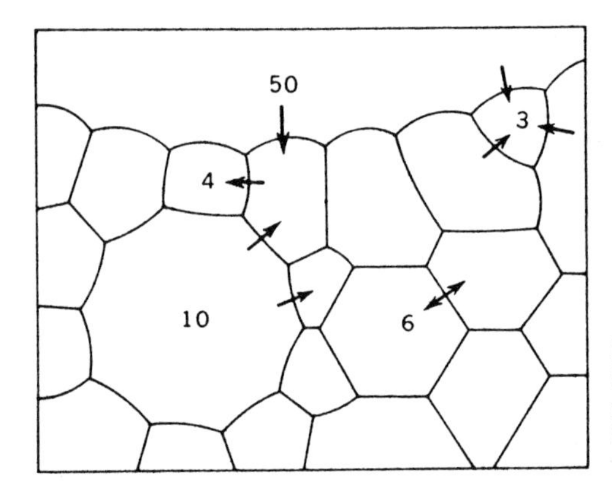

Figure 3.14 An array of grains showing the likely directions of boundary migration. After Shewmon, P.G., *Transformations in Metals*, McGraw-Hill, 1969 (quoting work of J.E. Burke)

In most practical steels, the presence of carbide or nitride particles tends to hinder grain growth, as illustrated in *Figure 3.15* (*a*). In this case the driving force for a migrating boundary, $2\gamma_{gb}/\bar{d}$, is counterbalanced by the increase in boundary area due to its interaction with the particles. If there is a volume fraction, V_f, of particles with radius r, the average number of interacting particles is $3V_f/2\pi r^2$, so that the restraining force, P, per unit area of boundary is given by equation (3.25). Equating P to the driving force for boundary

$$P \approx \frac{3V_f}{2\pi r^2} \pi r \gamma_{gb} = \frac{3V_f \gamma_{gb}}{2r} \tag{3.25}$$

migration, $2\gamma_{gb}/\bar{d}$, gives the equilibrium condition (3.26) when the particles

$$\frac{2\gamma_{gb}}{\bar{d}} = \frac{3V_f\,\gamma_{gb}}{2r} \tag{3.26}$$

prevent grain growth. For a given particle dispersion, therefore, equation (3.27) gives the maximum grain size. This relationship, first derived by

$$\bar{d}_{max} = \frac{4r}{3V_f} \tag{3.27}$$

Zener[14], is illustrated schematically in *Figure 3.15* (*c*), which shows that grain growth ceases after a certain time in the presence of a fine array of particles. An

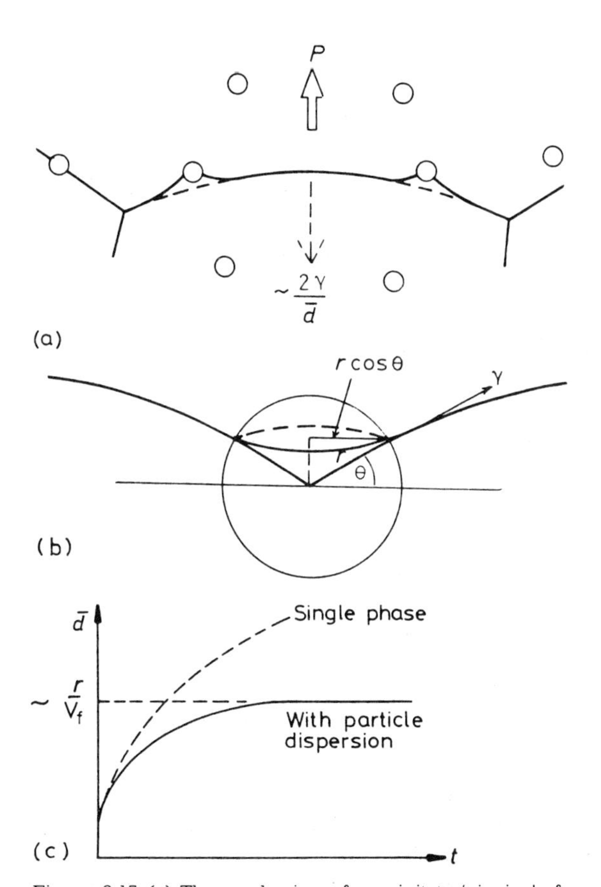

Figure 3.15 (a) The mechanism of precipitate 'pinning' of migrating grain boundaries. The detailed reaction between precipitate and boundary is controlled by surface energy, as shown in (b). In (c) the effect of precipitates on grain growth is shown. After Porter, D.A. and Easterling, K.E., *Phase Transformations in Metals and Alloys,* Van Nostrand Reinhold, 1981

alternative approach has been made by Gladman[15] which takes into account the distribution of grain sizes invariably present in practical materials. In this case equation (3.28) is derived, where z gives the ratio of the growing grain to

$$\bar{d} = \frac{\pi r}{3V_f}\left(\frac{3}{2} - \frac{2}{z}\right) \tag{3.28}$$

that of its neighbours, and during normal grain growth is expected to have a value between 1.5 and 2. The difference between the Zener and Gladman approaches can be understood in terms of an equilibrium and non-equilibrium situation. As illustrated in *Figure 3.15* (*c*), Zener's equation predicts the grain size after a certain time. Gladman's equation, on the other hand, essentially refers to the time at which grains are still growing. Note, from equations (3.27) and (3.28), the importance of particle size. In microalloyed steels, for example, typical carbonitride sizes are of the order of *ca.* 10—20 nm. This means that, typically, $V_f \ll 1$, giving normalized grain sizes of *ca.* 6—7 μm. Indeed, it is fairly hard to apply these equations in practice because of the difficulty in estimating V_f accurately. Another factor which should be noted in the Zener and Gladman approaches is that the grain sizes estimated essentially apply to a two-dimensional array. To obtain the three-dimensional grain size, a multiplication factor of 1.776 should be used; this factor relates the random intersect length through a two-dimensional array of grains with the three-dimensional grain size, assuming the grains have the shape of a tetrakaidecahedron.

According to equations (3.27) and (3.28), grain growth continues unhindered if the particles become dissolved. However, in practice certain elements, e.g. Nb, P and S, tend to segregate strongly at the austenite grain boundaries, and this exerts additional dragging forces. It is difficult to estimate the effect of impurity drag, particularly in a weld thermal cycle where the temperature — and hence the amount of impurities in grain boundaries — changes continuously. For an interesting discussion of this effect in welding, *see*, e.g., ref. 16.

Grain growth during welding, assuming particle dissolution

The problem here is to estimate the amount of grain growth in the HAZ under the essentially non-isothermal conditions of a thermal cycle. At the same time, the possible influences of particle pinning and impurity drag should be accounted for if the model is to be realistic. Treatments of this problem have been reported, e.g. *see* refs 16 and 17, but in view of the more general treatment given to this problem by ref. 11, it seems more appropriate to adopt the model of ref. 11 here.

Experimental observations have indicated that grain growth in many steels occurs predominantly at temperatures above the equilibrium solubility limits of carbides and nitrides. On this basis, it would seem appropriate to calculate grain growth over that part of the weld thermal cycle which exceeds the solubility temperature of carbonitrides, as illustrated in *Figure 3.16*.

The temperature–time profiles in the HAZ are obtained from the Rosenthal equations (Chapter 1). Only the simplified equation need be used since the portion of the thermal cycle affected by grain growth is unlikely to be influenced by, e.g., phase transformations. This, at least, gives a reasonable first approximation.

It is assumed that grain growth is diffusion-controlled, driven by surface energy and requires no nucleation. The extent of boundary movement thus depends on the integrated number of diffusive jumps during the weld thermal cycle. Thus, the rate of change of grain size, d, is given by equation (3.29),

$$\frac{\mathrm{d}d}{\mathrm{d}t} = A\, f(d) \exp - \left(\frac{Q}{RT(t)}\right) \tag{3.29}$$

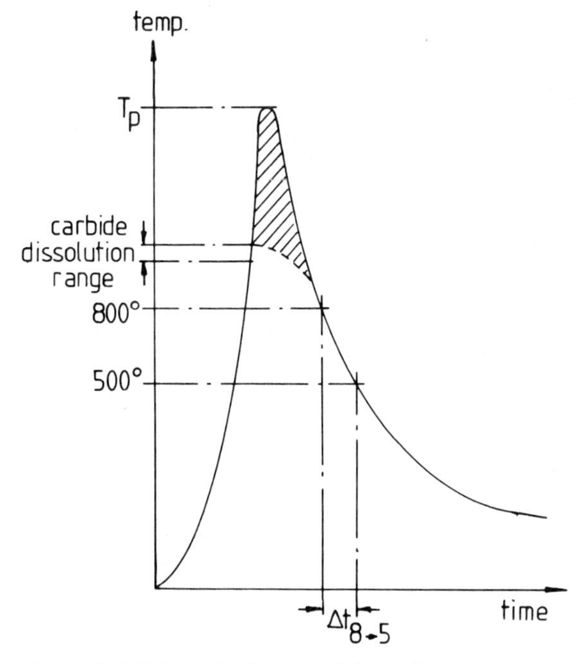

Figure 3.16 Schematic diagram of the weld thermal cycle. The shaded portion represents the part of the cycle in which unrestricted grain growth can occur, i.e. at temperatures above which the carbides are in solution. Experimentally it has been observed that most grain growth occurs during the heating part of the thermal cycle, *see* ref. 17

where A is a kinetic constant and $f(d)$ is any function of d. To obtain the total change in d over the thermal cycle, equation (3.29) is integrated as a function of time and temperature to give equation (3.30), where d_f and d_i refer to the final

$$\int_{d_i}^{d_f} \frac{\mathrm{d}d}{Af(d)} = I \tag{3.30}$$

and initial grain sizes. The left hand side of the equation depends only on d_i and d_f and, for a fixed change in d, it is a constant, I. Equation (3.31) thus follows

$$I = \int_0^\infty \exp - \left(\frac{Q}{RT\,(t)}\right) \mathrm{d}t \tag{3.31}$$

from equation (3.29). In effect, this gives the total number of diffusive jumps over the thermal cycle. The activation energy for self diffusion, Q, is dependent on impurity drag and has to be estimated from experiments. This can be done by fitting the contours of grain size as given by equation (3.30) to experimental data of grain growth as a function of T and Δt_{8-5}, as obtained from a weld simulator or from actual welds. The axes used in these grain-size diagrams are peak temperature and Δt_{8-5} or input energy*. The range of input energies is

TABLE 3.2 Relation between welding process and energy input

Welding process	*Approximate energy-input range ($MJ\,m^{-1}$)*
Electroslag	5–50
Submerged arc	1.0–10
Gas–metal arc	0.5–3
Manual metal arc	0.5–3
Gas–tungsten arc	0.3–1.5
Electron beam	0.1–0.6
Laser beam	0.1–0.6

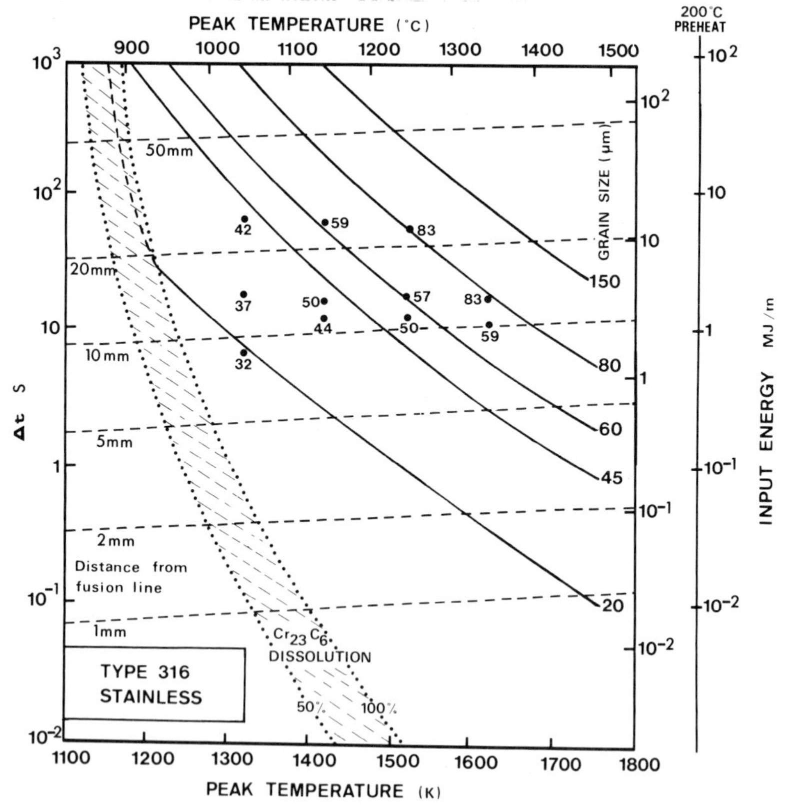

Figure 3.17 Austenite grain growth diagram for a bead-on-plate weld of stainless steel for $T_0 = 20$ and 200 °C. The dots represent experimental measurements of grain size. The carbide dissolution curves are adopted from *Figure 3.13*. The approximately horizontal dotted lines refer to width of the HAZ from the fusion line (without preheat). After Ashby and Easterling[11]

*Diagrams can also be drawn with axes of input energy and distance from weld centre line, *see Figure 3.34*

chosen to cover all the possible fusion-welding processes, as given in *Table 3.2*. To illustrate this approach two case studies are given below.

Case 1 Grain growth in an austenitic stainless steel The steel was a 19% Cr/ 11% Ni stainless of the type used, e.g., in corrosive environments, the chemical industry and in power plants. Samples for weld simulation were machined from cold-rolled plate, and the mean initial grain size, d, was estimated as *ca.* 18 μm. Samples were simulated assuming the thick plate condition over a range of peak temperatures (T_p) and cooling times (Δt_{8-5}). The slopes of the grain size contours obtained by computing equation (3.31) were then *adjusted* to fit the experimental points as well as possible, thus effectively accounting for 'drag-effects' due to alloying. The resulting diagram of grain growth is shown in *Figure 3.17*. For completeness, the dissolution curves for $Cr_{23}C_6$ carbides are included, the shaded portion defining the boundaries between 50% and 100% dissolution, i.e. 0.5 f and 1.0 f in equation (3.20). The full curves show the amount of grain growth for a given $\Delta t/T_p$ and the experimental results are marked as dots. The agreement between experiment and theory appears to be very satisfactory. The fine dotted lines on the diagram refer to distance from the fusion line as calculated from the Rosenthal

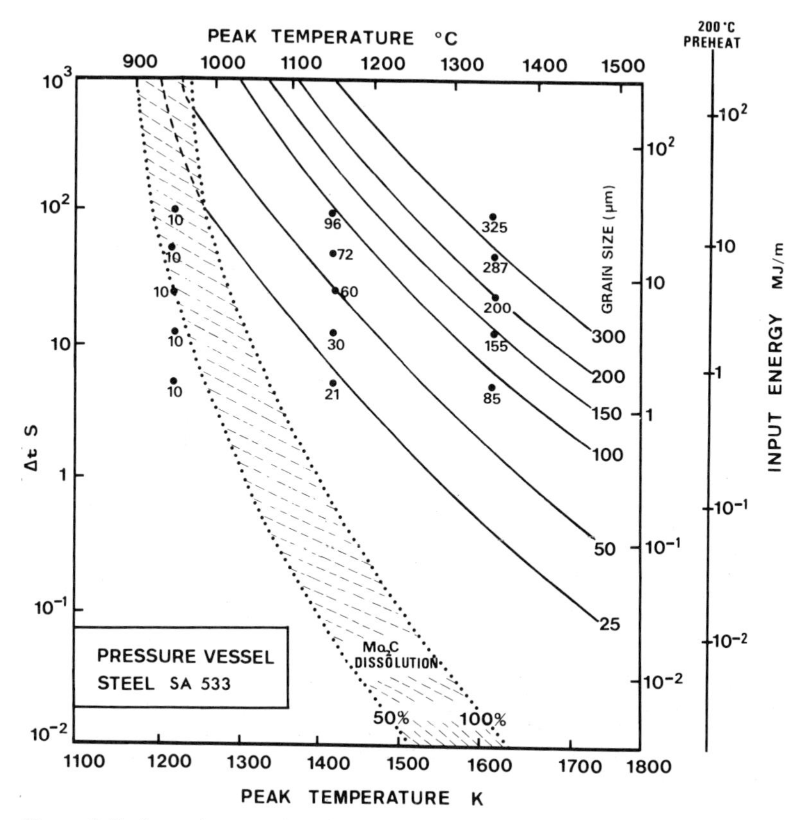

Figure 3.18 Austenite grain growth diagram for a bead-on-plate weld of a pressure vessel (ferritic–pearlitic) steel; T_0 = 20 and 200°C. After Ashby and Easterling[11]

equations. This thus provides a means of estimating the width of the grain growth zone. Assuming the edge of the grain growth zone corresponds to the carbide solubility boundary, then for, say, a heat input of 1 MJ m^{-1}, $T_0 = 20$ °C, the total width of the grain growth zone is predicted to be *ca.* 7.5 mm for a bead-on-thick plate weld. The maximum grain size at this heat input is then *ca.* 70 μm.

The additional input-energy scale to the right of the diagram refers to the imposition of preheat prior to welding. Since the energy from the preheat treatment is additive to the welding input-energy, the net effect is to *increase* grain growth. At the same time preheat also increases Δt_{8-5}, which reduces the possibility of martensite formation.

Case 2 Grain growth in a pressure vessel steel Pressure vessel alloys are based on low alloy, high strength steels that have good weldability and ductility. They typically have a yield strength of around 345 MPa and a tensile strength of *ca.* 600 MPa with a ductility of *ca.* 18 %. The steel chosen for the diagram contains 0.25 wt % C; 1.2 wt % Mn; 0.5 wt % Mo, and has a mixed microstructure of ferrite and carbides (tempered bainite). The steel is used in applications such as bridges, ship plate and pressure vessels. The welding diagram for the steel is shown in *Figure 3.18*. The most stable carbides are expected to be those based on Mo as shown. However, since these carbides dissolve at relatively low temperatures, considerable grain growth results in this case, particularly in the high input-energy range. As with stainless steel the agreement between theory and experiment appears satisfactory.

Grain growth during welding, assuming particle coarsening

This represents a somewhat more complex situation than that considered in the previous section. It is also an important case to consider since the obvious way to control grain growth in high energy welds is to develop alloys which contain particles that remain stable during the weld thermal cycle. An interesting practical example of such a steel is the Ti-microalloyed quality based on dispersions of TiN precipitates.

To estimate the amount of grain growth in the HAZ, the fact that during the weld thermal cycle the mean particle size increases while the volume fraction decreases has to be taken into account. Therefore equation (3.15) has to be used to obtain the particle coarsening as a function of weld thermal cycle (T_p and Δt_{8-5}). The values obtained can then be used in the Gladman equation (3.28), after making suitable adjustments for changes in volume fraction as derived from equation (3.19). To illustrate this procedure, an example from the current research program at the University of Luleå[18] is given.

Case 3 Grain growth in a Ti-microalloyed steel The Ti-microalloyed steel used in this study is one of a number of experimental steels prepared by the Swedish Steel Corporation to study the stability of TiN and other mixed carbonitrides. In this case, the steel contains only TiN precipitates and coarsening is studied after various simulated thermal cycles, using carbon extraction replicas and transmission electron microscopy (TEM).

As emphasized earlier, the greatest difficulty in applying the Zener–Gladman equations is to estimate the volume fraction of particles, V_f. The problem is that V_f cannot be estimated merely on the basis of the steel's known composition, since the formation of coarse TiN inclusions cannot easily be avoided during processing. It may not be accurate, either, to estimate V_f from extraction replicas or even from TEM studies of thin sections. An indirect

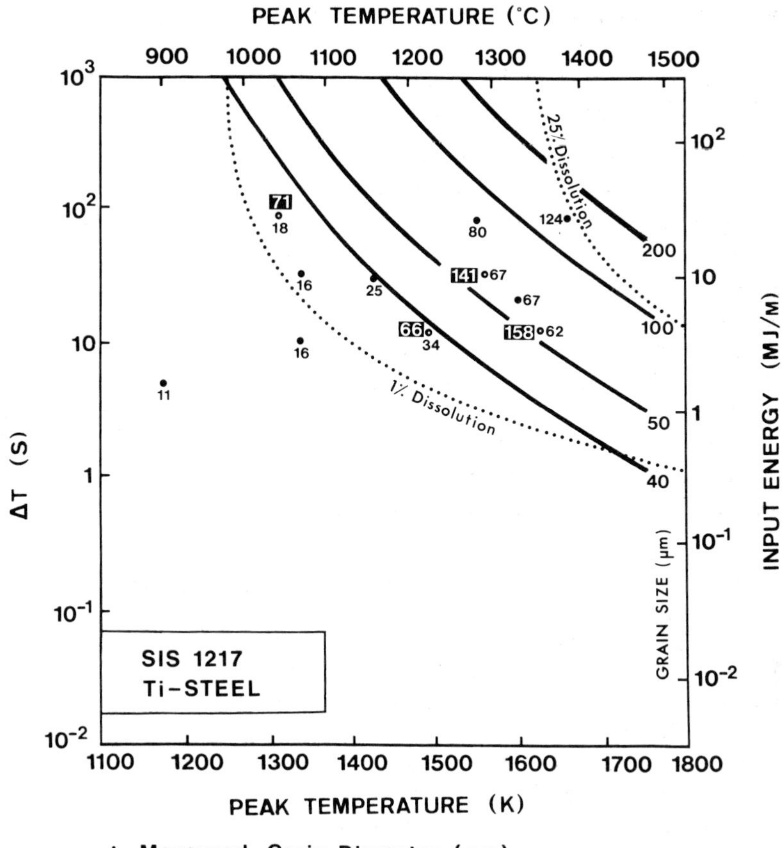

Figure 3.19 Plot of austenite grain growth in the HAZ of a Ti-microalloyed steel based on a particle coarsening model (*see* text). The points given in the diagram are from experimental measurements. The precipitate in this case is TiN. After unpublished research of J. Ion and K.E. Easterling, University of Luleå

approach is to assume that at a given time the Gladman equation (3.28) is valid, so that knowing the grain size, d, and the particle size, r, V_f can be estimated accordingly.

Hence, from equation (3.28), assuming $z=2$, equation (3.32) is obtained,

$$d \approx 0.5\left(\frac{r}{V_f}\right) \tag{3.32}$$

which gives the two-dimensional grain diameter. The changes in V_f as a function of T_p and Δt_{8-5} are obtained from equation (3.19), using a Q_v/RT_m value of 14.4 for the diffusion of Ti in austenite. From this, the original volume fraction is estimated to be 1.235×10^{-4}. Using these values, the grain sizes can be calculated and assembled diagramatically, as shown in *Figure 3.19.* As before, the data are assembled in terms of welding input-energy (or Δt) and peak temperature. The TiN dissolution lines (dotted) are shown for dissolution of 1% and 25%. The boxed numbers are the measured precipitate diameters (compare with *Figure 3.11*). As in the previous diagrams, the full lines are the contours of grain size. Considering the doubts about the estimation of V_f, there appears to be fairly reasonable agreement between the experimental measurements of grain size and the estimated sizes (contours). Significantly, it is seen that even in the presence of relatively stable TiN particles, grain growth in the HAZ at high input-energies can be quite considerable. On the other hand, compared with pressure vessel steel (*Figure 3.18*), grain growth is estimated to be only about a third as much.

As discussed above, it is quite difficult to give a good estimation of V_f on the basis of electron micrograph replica observations. A better approach might be to estimate the mean distance between particles that actually intercept grain boundaries from thin foil micrographs, and relate this with $\bar{d}$. This is particularly relevant to many fine grained microalloyed steels in which only a few particles present are in contact with the boundary, *see*, e.g., *Figure 3.2*.

Practical considerations of grain growth and grain growth control in the HAZ

It can be concluded from the previous sections that a reasonable correlation between theoretical prediction of grain growth in a weld thermal cycle and that obtained experimentally using weld simulated specimens is possible. Of course, in real welds the geometry of the weld may not always give a uniform HAZ width and this is illustrated in *Figure 3.20,* for a bead-on-thick plate Nb-microalloyed steel weld. It is seen that because of problems of penetration, a non-uniform width of the HAZ has developed and exhibits variations in maximum grain size. Variations such as this have to be taken into account when applying the grain growth diagrams to real welds.

Grain sizes in the HAZ of real welds can be correlated with temperature by applying the Rosenthal equation between fixed boundary conditions, say the fusion line (1800 K) and the cementite coarsening–recrystallization transition (*ca.* 1000 K), to obtain the temperature distribution as a function of microstructure. Examples of this for MIG and SA welded Nb-steel are illustrated in *Figures 3.21* (*a*) and (*b*). In both cases, the average grain sizes were measured along a number of isotherms and fairly good agreement between theory and experiment was obtained. However, the difference in maximum grain size between the two welding techniques is due to the weld geometry, it being symmetrical for the case of the SA weld, and non-symmetrical for the MIG weld.

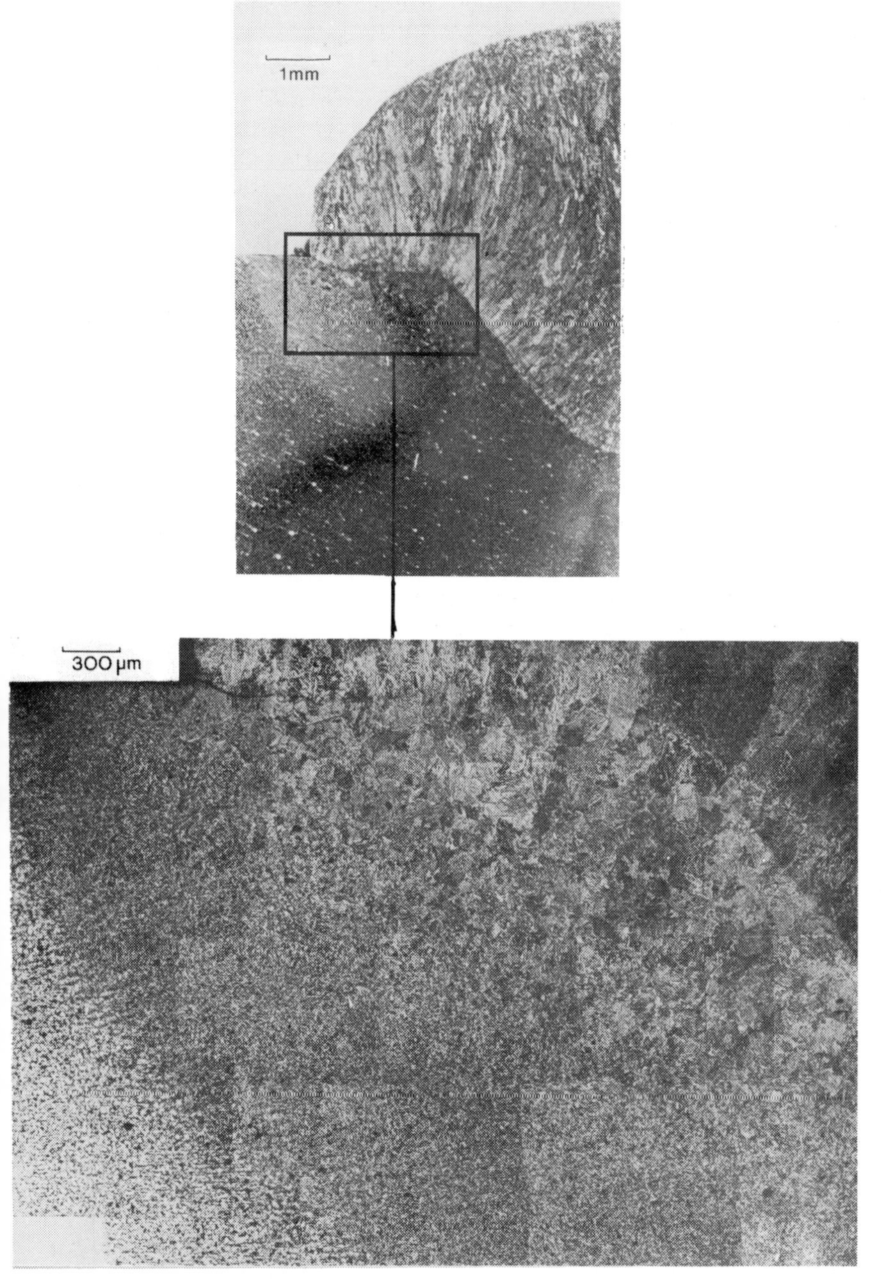

Figure 3.20 A bead-on-plate MIG weld of a Nb-microalloyed steel. Note that due to high penetration of the weld at the centre, the HAZ width is not uniform and most grain growth has occurred at the concave section of the weld. *See also Figure 1.27.* Courtesy of J. Strid and J. Ion, University of Luleå

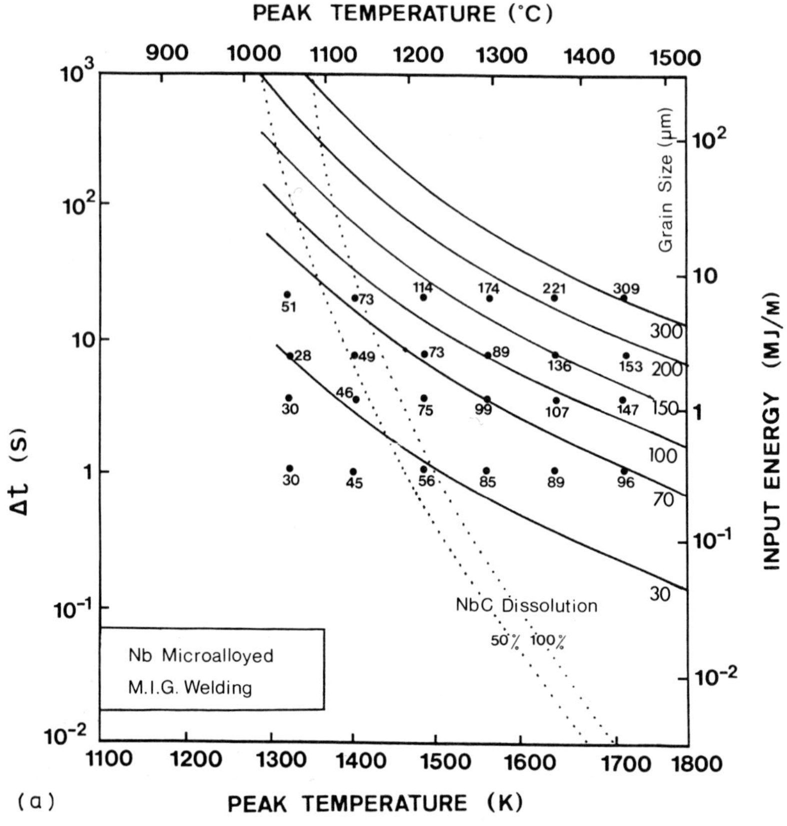

Figure 3.21 (a) Grain growth diagram (thin plate condition) for a Nb-microalloyed steel. The weld geometry is assymmetrical like that shown in *Figure 3.20*. The points refer to the mean grain sizes around actual MIG welds. The grain-size contours are computed assuming $Q/RT_m = 13$. After unpublished work of J. Ion and K.E. Easterling, University of Luleå

It is interesting that the maximum increase in grain size occurs in the initial stages of grain growth, because of the higher driving force available at this stage. This has been confirmed, e.g., in ref. 17, in which grain growth in TIG- and SA-welded microalloyed high strength steels is discussed. This work shows that the cooling cycle only accounts for about 20% of the total grain growth, which again emphasizes the importance of using a correct heating cycle when employing the weld simulator.

In comparing HAZ grain growth in weld simulated and real welds it is sometimes found that the maximum grain size in real welds is less than that in simulated welds. This observation appears to apply mainly to transformable steel welds in the medium heat input range. Assuming that the correct heating cycle is employed, a possible reason that has been suggested for this phenomenon is that grain growth of the large grains in real welds is hindered by the smaller grains, rather like a 'pinning' effect. The situation is illustrated in *Figure 3.22*, in which it is seen that the change in grain size is associated with a very steep temperature gradient. This gradient is particularly severe in medium

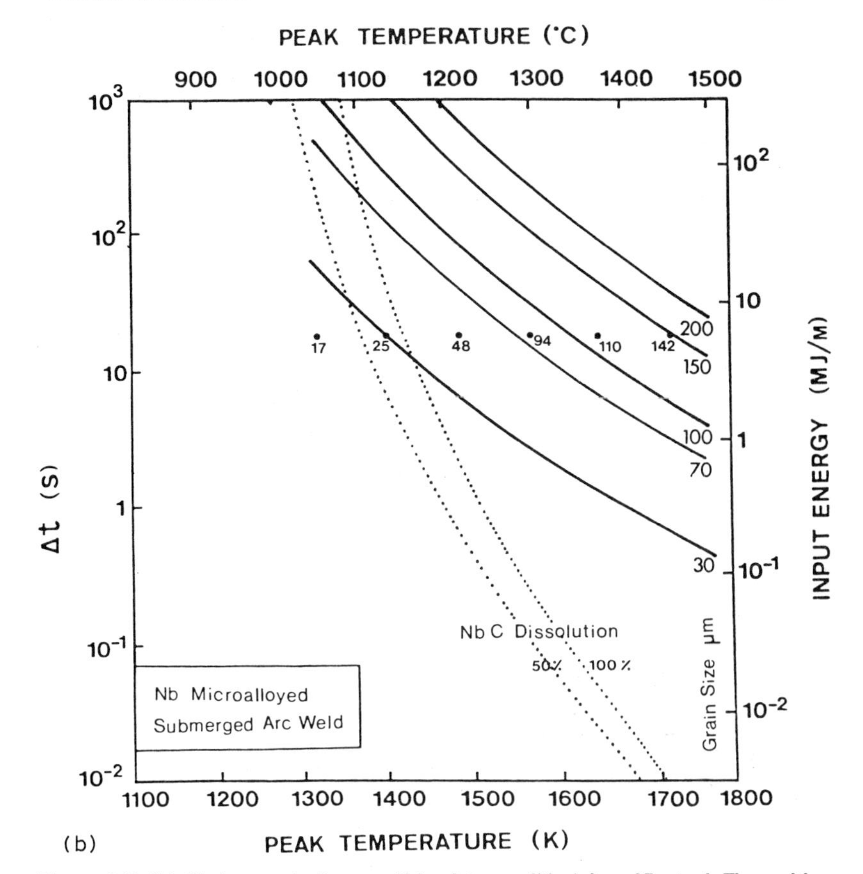

Figure 3.21 (b) Grain growth diagram (thin plate condition) for a Nb-steel. The weld geometry in this case is quite symmetrical. The points refer to mean grain sizes around actual SA welds. As in (a), the contours of grain size are computed assuming $Q/RT_m = 13$. After unpublished work of J. Ion and K.E. Easterling, University of Luleå

heat input steel welds because of the high melting point and relatively poor conductivity of steel. The origin and magnitude of such a pinning effect is difficult to quantify. One possibility is that grains which have a large temperature gradient across them tend to grow non-uniformly, resulting in a change of shape from, say, equiaxed to pear-shaped. The corresponding change in surface-to-volume ratio effectively represents an increase in energy, thereby neutralizing the reduction in energy present in normal grain growth. There could also be surface tension restrictions experienced by grains when adjacent grains are trying to expand at faster or slower rates.

It is apparent from the above considerations that some degree of grain growth control can be exerted, either by limiting the duration of the weld thermal cycle as, e.g., in low energy-input welding processes, or by precipitate pinning. Since many other factors determine the type of weld process, or weld thermal cycle employed, most research in recent years has concentrated on improving the weldability of steels subject to high energy welding processes, by means of microalloying.

It is seen from *Figure 3.13* that the most stable precipitate in steel is TiN, which only begins to go into solution at the highest welding energies. Work on carbide and nitride solubilities has shown, in fact, that steels can be produced with dispersions of TiN precipitates of a size, dispersion and volume fraction sufficient to hinder grain growth at least up to temperatures of *ca.* 1200—

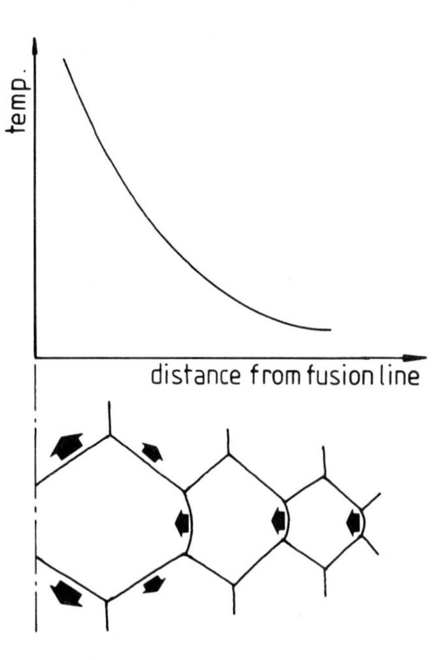

Figure 3.22 Grain size in the HAZ as a function of peak temperature and distance from the fusion line. The rapid change in grain size may tend to hinder grain growth, e.g. because of grain shape changes as shown

1250 °C [19]. As a general rule, the simple relationship between grain size, precipitate size and volume fraction (equation (3.27)) is in reasonable agreement with experimental results on commercial TiN-steels. For example, in a detailed study by Matsuda and Okumura[20] the *empirical* relationship (3.33) was derived from experiment, in reasonable agreement with equation

$$\bar{d} = 0.75 \frac{r}{V_f} \tag{3.33}$$

(3.27). An example of these results is shown in *Figure 3.23*. On the other hand, Kanazawa[21] studied a relatively pure Ti-steel and determined a value of 0.24 for the constant of proportionality, equation (3.34). It is interesting to note that

$$\bar{d} = 0.24 \frac{r}{V_f} \tag{3.34}$$

the constant of proportionality for the Gladman equation (assuming $z = 2$), lies roughly between those given by equations (3.33) and (3.34).

As a general rule, the mean distance between particles, $\bar{x}$, is given approximately by the ratio r/V_f (provided $V_f \ll 1$), equation (3.35), which

$$\bar{x} \approx \frac{r}{V_f} \tag{3.35}$$

confirms, once again, that the best way to control grain size is to use large amounts of very small precipitates. Unfortunately, in practice it appears to be difficult to achieve this by the use of TiN particles alone. For example, it seems that the optimum amounts of Ti and N that can combine to precipitate as TiN are *ca.* 0.03—0.06 wt % Ti and *ca.* 0.008—0.017 wt % N [22]. Higher amounts of Ti have two negative effects, firstly causing large, coarsely spaced inclusions of TiN to form, and secondly increasing the tendency for the embrittling TiC precipitate to form in addition to TiN[19,22].

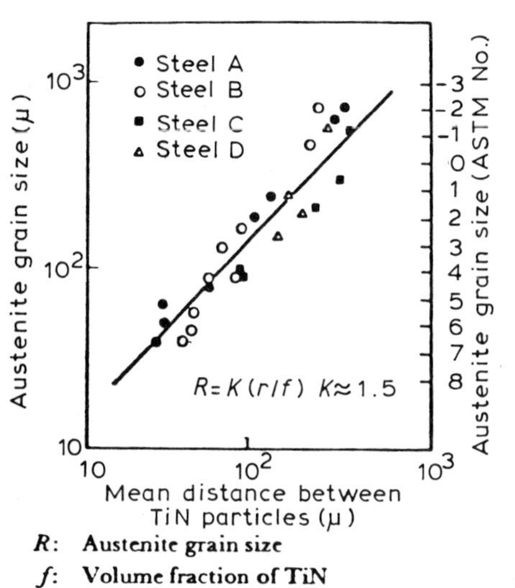

R: Austenite grain size
f: Volume fraction of TiN
r: Mean particle size of TiN
K: Dimensionless constant

Figure 3.23 Relation between mean spacing of TiN precipitates and austenitic grain size. After Matsuda and Okumura[20]

In order to meet certain minimum requirements of yield strength, steel producers find it necessary to add Nb and/or V in addition to Ti in order to provide the required degree of grain refinement. Thus, Ti-bearing steels may contain a mixture of various carbonitrides, including TiN, VN, AlN, and even NbC. During a high energy weld cycle, only the TiN is likely to survive the peak temperatures and coarsening even of these precipitates is inevitable. Recent thin foil microanalysis work suggests that, on both heating and cooling of the weld, the undissolved TiN particles act as reprecipitation sites for, e.g., V and Al that dissolved during heating. Thus, at ambient temperature the particles which remain are relatively coarse and have a composition of the type (TiVAl)(N). This sequence of dissolution, coarsening and reprecipitation is illustrated schematically in *Figure 3.24.* It appears that the reprecipitating elements do not always mix homogeneously with the TiN, but instead may sometimes form shells around the original TiN particles[23]. This may even be expected on the basis that, for example, TiN and AlN are not isomorphous with one another. As a result of this coarsening and mixing process, in subsequent thermal cycles the mixed particles do not offer as much resistance to grain growth in the higher temperature ranges, and hence these steels do not retain their small grain size properties during reheating cycles. It appears that mixed nitrides are not as stable as simple TiN, so that the solubilitiy temperatures are lowered to below those predicted, e.g., in *Figure 3.13*.

TiN has been discussed mainly as a grain refiner of the HAZ. According to *Figure 3.10* other particles of comparative thermodynamic stability are ZrN and HfN. Indeed, there is some evidence that these precipitates can also be effective in reducing grain growth[22]. The borides ZrB_2 and HfB_2 may also be

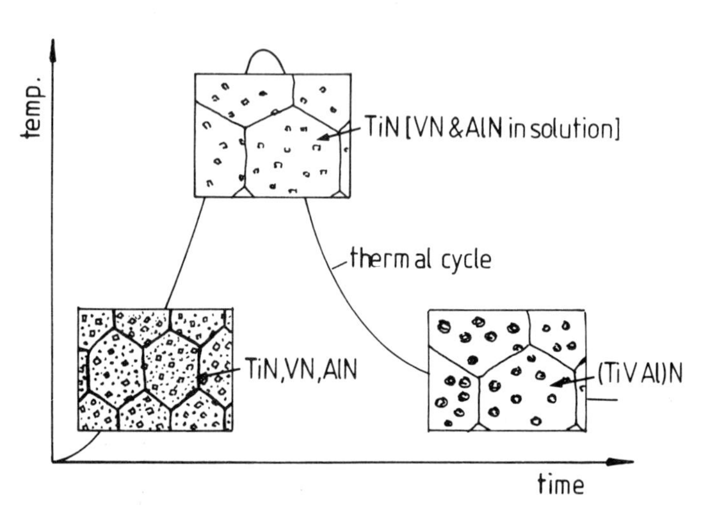

Figure 3.24 A schematic illustration of possible precipitate dissolution, coarsening and reprecipitation in a Ti–V–Al microalloyed steel, as based on research at the University of Luleå

worth considering as grain-growth inhibitors in B-steels. Alternatively, rapid weld thermal cycles (short heating times) do not allow complete dissolution (*Figure 3.13*) and this affects grain growth accordingly.

Reactions at the fusion line

Reference to *Figure 3.20* shows that the grain size decreases slightly at the fusion line in spite of the higher temperature at the fusion line. In steel, this may depend on the $\gamma \rightarrow (\delta + \gamma)$ phase transformations, as pointed out in ref. 10. In order to clarify the various reactions and microstructures at the fusion line, consider the phase diagrams and microstructures shown in *Figure 3.25*. In case (*a*), the alloy composition is such that the base metal at the fusion line transforms to δ-ferrite and liquid at the grain boundaries. Because of the lower solubility of C and Mn in ferrite (both austenite formers) these elements tend to segregate to the melted grain boundaries. On transforming back to austenite during the cooling cycle, the high degree of segregation at the boundaries increases boundary drag and thus reduces grain growth. It also very effectively modifies the composition of the austenite, and this, together with the effect of a smaller grain size, affects the austenite transformation products, compared with the larger austenite grains away from the fusion line, by transforming at higher temperatures. This accounts for the more equiaxed ferritic structure at the fusion line.

In case (*b*), which is typical of plain C–Mn steels, the finer grain size at the fusion line is not observed. As seen in *Figure 3.25* (*b*), there is no $\gamma \rightarrow (\delta + \gamma)$ transformation in this case, so that austenite transforms only to (γ + liquid) at the grain boundaries. Since there are fewer segregation problems in this case, the composition of the austenite at the fusion line is not expected to change

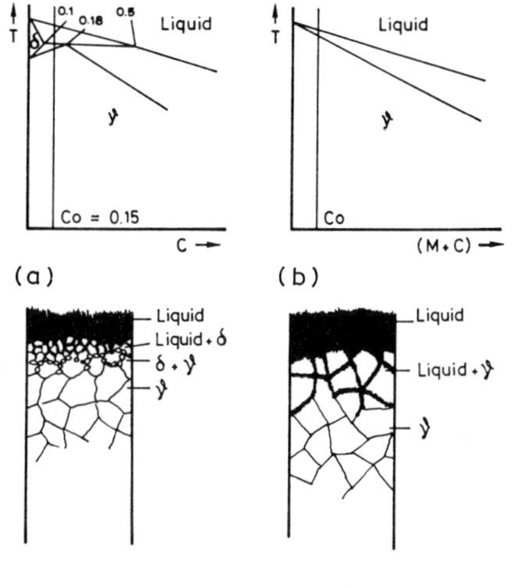

Figure 3.25 Schematic illustration of microstructures of steels at the fusion line (a) with and (b) without the $\gamma \leftrightharpoons \delta$ transformation occurring. After Räsänen and Tenkula[10]

substantially, and transformation during cooling of the unmelted part of the grains occurs at similar temperatures to those grains not in contact with a liquid phase.

It may be conjectured that during heating in the austenite range impurities that segregate to the grain boundaries may, in some cases, produce compounds, e.g. (MnFe)S, which lower the melting point of the boundary regions. Thus, in high sulphur steels, it is expected that the grain boundary melted region may be wider. During cooling, these low melting point films persist, and as residual stresses increase at the lower temperatures, cause *liquation cracking,* as illustrated in *Figure 3.26.* Besides steels, this phenomenon is known to occur also in nickel-based and aluminium alloys. The tendency for segregation to grain boundaries can be estimated according to how soluble the elements are in a given phase. On this basis a grain boundary enrichment factor, β_b, is defined by equation (3.36) where C_b is the equilibrium

$$\beta_b = \frac{C_b}{C_o} \tag{3.36}$$

concentration of the element in the boundary, and C_o is the equilibrium concentration in the matrix. β_b has been measured in a number of alloys using scanning Auger spectroscopy, and a summary of these results is given in *Figure 3.27.* The horizontal dotted line at $\beta_b \approx 100$ represents the approximate boundary between segregants that tend to give embrittling effects and those

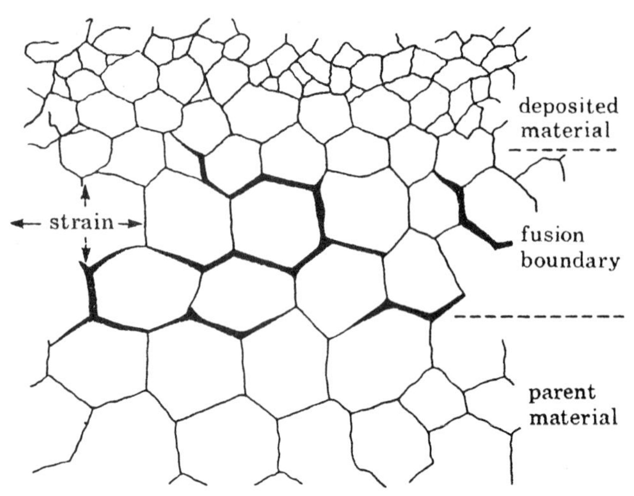

Figure 3.26 Schematic illustration of liquation cracking at the fusion line. After Baker, R.G., Weldability and its implications for materials requirements, *Rosenhain Centenary Conference Proceedings*, p.129, eds Baker, R.G. and Kelly, A., Royal Society, London, 1975

that do not. Thus, for values of $\beta_b > 100$, embrittlement is a possibility. This implies, for example, that Bi in copper is highly embrittling. In the case of iron alloys, the most embrittling elements are thus predicted to be S, C, B, P, N, Sb and Sn, in decreasing order of effectiveness.

A particularly important group of alloys in respect to embrittlement in the HAZ is the corrosion-resistant ferritic stainless steels, since grain boundary segregation can decrease their resistance to intergranular corrosion. This

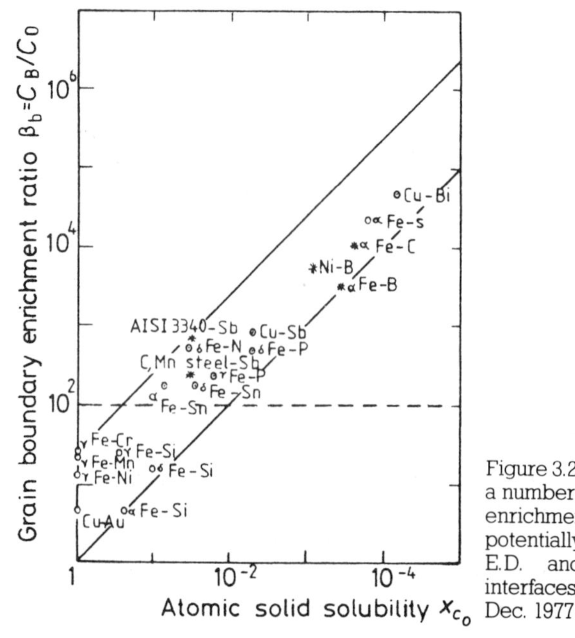

Figure 3.27 The effect of the solid solubility of a number of alloy systems on grain boundary enrichment. Values of $\beta_b \gtrsim 100$ depict potentially brittle materials. After Hondros, E.D. and Seah, M.P., Segregation to interfaces, *International Metals Review*, **222**, Dec. 1977

problem is associated with the $\alpha \rightarrow \gamma$ transformation in the grain growth zone which occurs at grain boundaries. Intergranular austenite becomes impoverished in Cr and enriched in C, and during rapid cooling of the weld these regions transform to martensite, or tempered martensite that contains carbides. Either because of the Cr-depleted martensite or of the presence of the Cr-carbides, these transformed regions degrade the usually good stress-corrosion resistance of the steel.

It should be noted, finally, that the segregation arguments that concern grain boundary embrittlement apply equally well to reheating of welds, and in some cases susceptibility may be higher because of the lower solubility of these elements in the matrix at the lower temperatures used in, e.g., stress relief annealing.

Transformations during cooling

As discussed in Chapter 2 (pp. 75–83), the type and volume fraction of transformation product depends on several factors, including grain size, peak temperature, cooling rate, alloying elements present, etc. However, the high carbon equivalents of the base metal compared to the weld metal tend to move the C-curves of CCT diagrams to longer times, thereby increasing the risk of martensite formation. In high energy welds, a high C_{equiv} tends to increase the amount of pearlite formed. In low C_{equiv} steels, on the other hand, an embrittling structure known as *interphase precipitation* may occur. This consists of dense rows of fine carbides which precipitate at the moving γ/α interface by a ledge mechanism. It is useful to briefly consider the formation of the various microstructures of the sub-zones in the HAZ, and for this purpose reference should be made to *Figure 3.28*.

Grain growth zone (*Figures 3.28* (*a*) and (*b*))

In the lower C_{equiv} steels, proeutectoid ferrite networks (white phase) which cover grain boundaries are a prominent feature, but higher C_{equiv} values tend to reduce this for lower temperature transformation products, particularly Widmanstätten side plates. Interphase precipitation is known to occur in low hardenability alloys, especially the Nb-bearing microalloyed steels, and the presence of this phase is likely to adversely affect toughness. Martensite forms in the higher C_{equiv} steels and, since it is the final product to form, it is mainly concentrated at mid-grains.

Grain refined zone (*Figure 3.28* (*c*))

The reduction in peak temperatures in this zone (*ca.* 1100 °C) implies that, following the $\alpha \rightarrow \gamma$ transformation during heating, the austenite does not have time to develop properly, and the grain size remains very small. In addition, carbides may not be fully dissolved. The $\gamma \rightarrow \alpha$ transformation on cooling, therefore, tends to produce a fine grained ferrite–pearlite structure

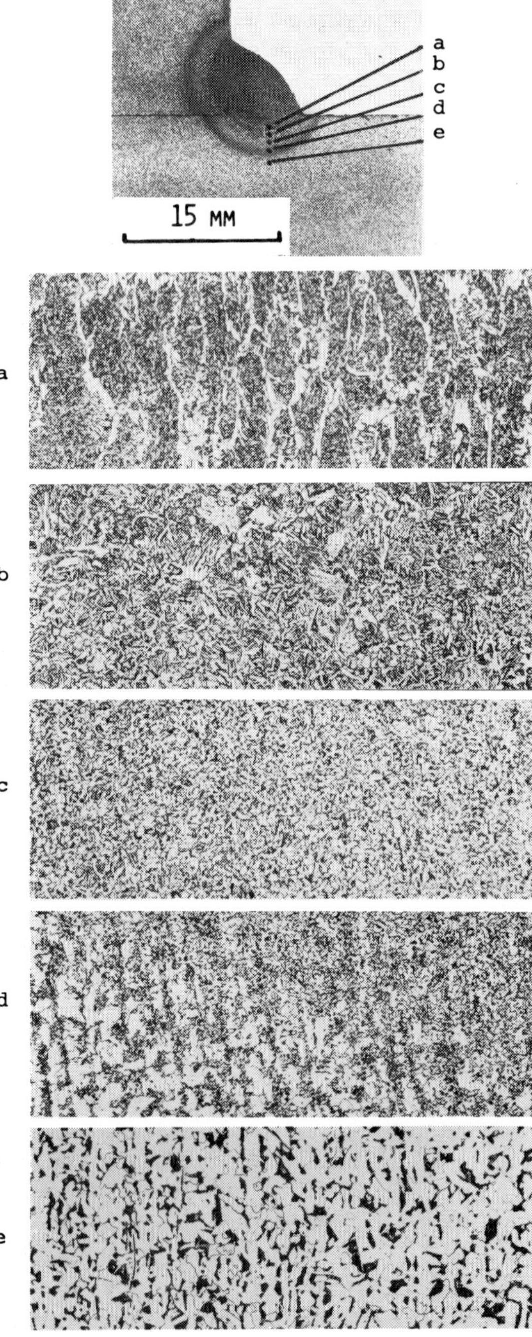

Figure 3.28 A series of optical micrographs showing the various zones in the HAZ of a MMA-welded normalized fine-grained pearlitic steel. (a) Fusion zone, (b) grain growth zone, (c) grain refined zone, (d) partially transformed zone and zone of spheroidized carbides, (e) unchanged base metal. Courtesy of Sotkovszki, P., Chalmers University of Technology, Gothenburg

depending, of course, on welding energy-input, plate thickness, etc. The large grain boundary area tends to promote ferrite nucleation and the austenite that remains at the grain centres is rich in carbon and transforms to pearlite. This zone tends to be particularly wide in the microalloyed steels, because of the effectiveness of the carbonitrides in preventing grain growth at these temperatures.

Partially transformed zone (*Figure 3.38* (*d*))

In the temperature range 750–900 °C, the pearlite in the base metal quickly austenitizes because of its lower $\alpha \rightarrow \gamma$ transformation temperature, the austenite being enriched in C and Mn. For this reason, the $\gamma \rightarrow \alpha$ transformation on cooling can yield a wide range of possible structures, depending on cooling rate, Δt_{8-5}, e.g. pearlite (dark phase), upper bainite, autotempered martensite or high carbon (twinned) martensite.

Zone of spheroidized carbides (*Figure 3.28* (*d*))

The spheroidized carbide zone corresponds approximately to the temperature range 700—750 °C. There is relatively little $\alpha \rightarrow \gamma$ transformation during the rapid heating cycle, so that the most notable change concerns degradation of the lamellar pearlite to spheroidal particles of Fe_3C (dark phase). The agglomeration of spheroidized cementite particles at grain boundaries and triple junctions, *Figure 3.28* (*d*), emphasizes the role of grain boundaries as high diffusivity channels for carbon at these low temperatures.

Zone of 'unchanged' base material (*Figure 3.28* (*e*))

The 'unchanged' zone concerns temperatures in the range up to *ca.* 650 °C, in which changes in morphology of constituents do not appear to occur. However, the combined effect of heating and residual stress can cause *dynamic strain ageing* to occur. This phenomenon is associated with moving dislocations sweeping up interstitial impurities such as C and N. On cooling, the solute enriched dislocations are strongly locked in position, thus embrittling the structure. The problem can be intensified by additional welding runs, as in multi-run welds, or during reheat annealing.

Predicting the microstructure and properties of the HAZ

The production of high strength structural materials is mainly based upon developing a product in which the grain size is as small as possible. The combination of good strength and toughness with low transition temperatures in steels is thus obtained by sophisticated treatments which refine the grain size without hardening the grains. The severity of the weld thermal cycle is such that this structure is completely modified near the weld. Indeed, in high energy processes such as submerged arc, it is not uncommon for the grain size to be increased tenfold or more in a zone close to the fusion line. In steels of high carbon equivalent ($\gtrsim 0.4$), it is likely that the grain growth zone contains martensite. Furthermore, the possibility of martensite formation increases at lower heat inputs as Δt_{8-5} decreases, and the same is true with larger grains.

The microstructure of the grain growth zone, above all other zones in the HAZ, determines the properties of the weld. The ability to make predictions of the microstructure and properties of this zone requires knowledge of the amount and extent of grain growth and of the weld thermal cycle. The extent or width of the grain growth zone is important in determining the maximum length of easy crack propagation in relation to fracture toughness. The thermal cycle is important in determining the extent of grain growth and the cooling rate. There are basically two approaches: first, to carry out large scale tests of the welded joint of interest, which is likely to be expensive; second, to try to make predictions based on small-scale experiments, e.g. weld simulations, Charpy V-notch, crack-opening displacement (COD), etc, or to use a theoretical model, or a combination of both. The latter approach is probably less reliable, but it is much less expensive. A few examples of the latter approach are considered below.

Weld simulation

The advantages and disadvantages of this technique have been discussed in Chapter 1. It is a very useful method for producing Charpy or COD test samples and microscopy specimens from any required part of the heat-affected zone. Even the effect of preheat and post-weld heat treatments, multi-run welds, etc, can be easily incorporated in the programmed thermal cycle. An example of this application of weld simulation in connection with a high strength steel is illustrated in *Figure 3.29* (*a*), in which the impact toughness and transition temperature of the various HAZ microstructure zones are compared with that of the parent metal. *Figure 3.29* (*b*) shows the effect of various post-weld anneals. These types of measurements enable prediction of the *likely* changes in properties of the real weld compared with the base plate and might suggest possible ways of improving the properties, e.g. by annealing.

In certain critical cases, e.g. where the real weld microstructure changes rapidly with the temperature gradients, weld simulation may not be a fully reliable method for producing test samples. For example, as mentioned earlier, the crack length in such a weld might be restricted in its propagation if the grain growth zone is very narrow. In this case, simulated samples which contain a uniform grain size that corresponds to the maximum grain size of the weld might exhibit lower toughness than the real weld.

Hardness measurements

It is fairly easy to make hardness measurements and an example of a hardness profile is shown in *Figure 3.30*. In spite of the changes in grain size the hardness profile is relatively uniform, increasing to a maximum at the fusion line, which gives a useful indication of the microstructural constituents and of the presence of martensite. By plotting the grain size as a function of hardness it is also easy to see the extent of the grain growth zone and its relation to the weld

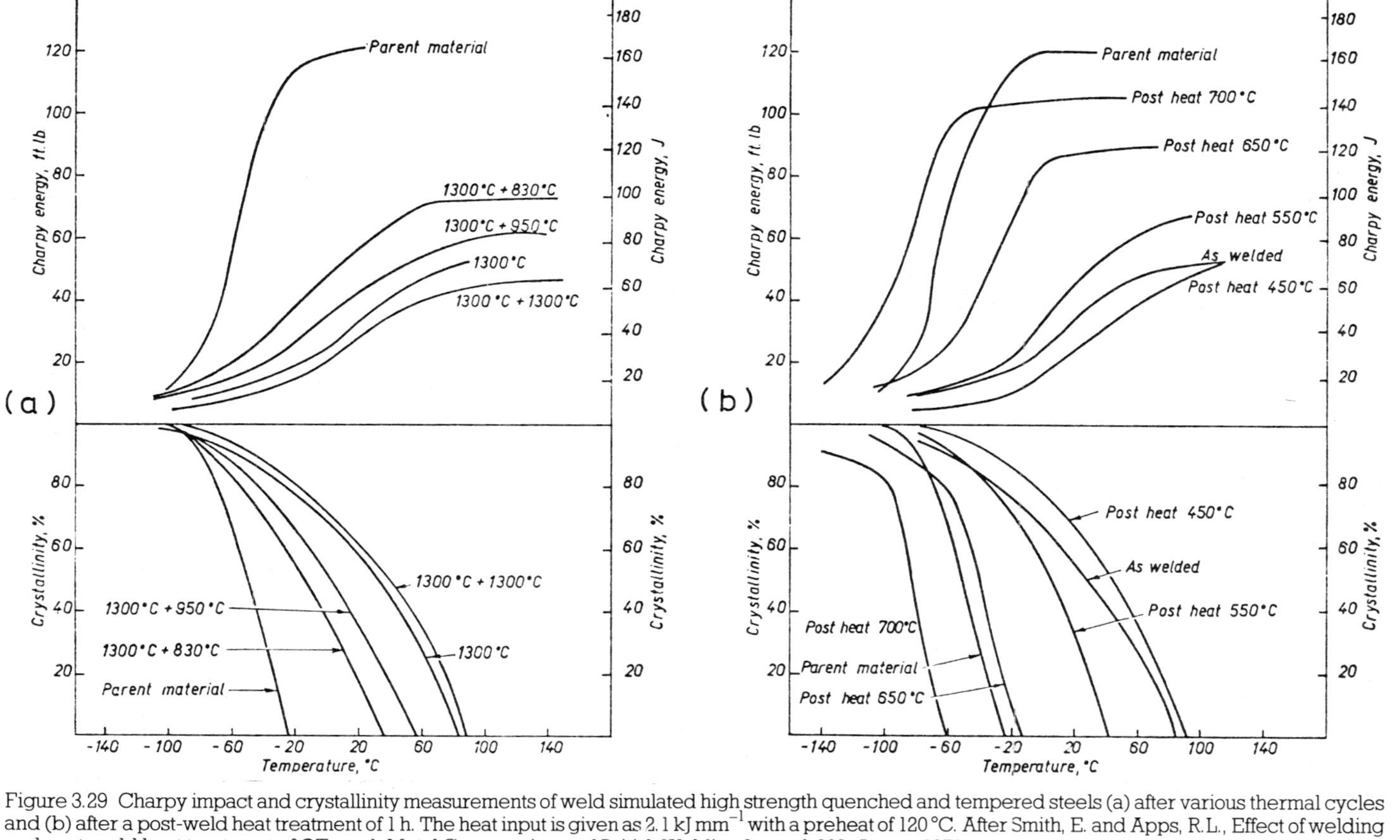

Figure 3.29 Charpy impact and crystallinity measurements of weld simulated high strength quenched and tempered steels (a) after various thermal cycles and (b) after a post-weld heat treatment of 1 h. The heat input is given as 2.1 kJ mm^{-1} with a preheat of 120 °C. After Smith, E. and Apps, R.L., Effect of welding and post-weld heat treatment of QT steel, *Metal Construction and British Welding Journal,* 303, August 1971

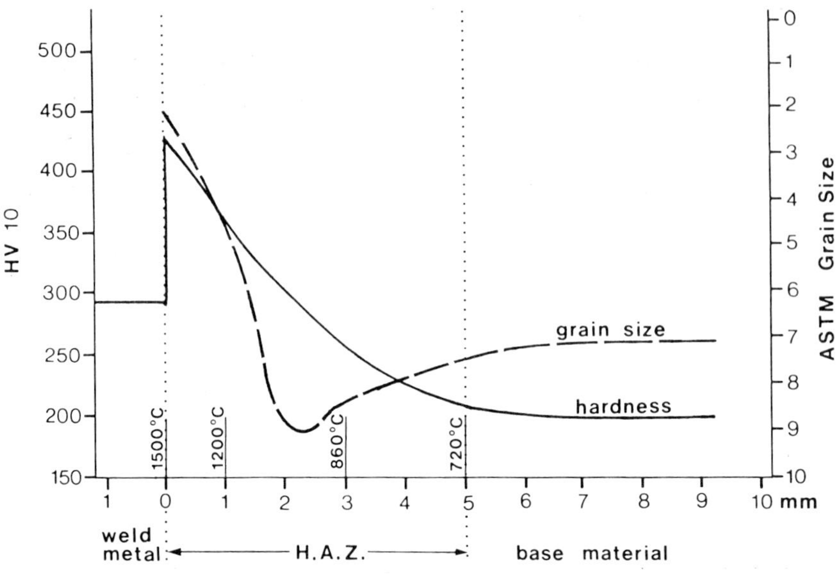

Figure 3.30 Vickers hardness measurements in the HAZ of a structural steel. From DVS Berichte No 65 Conference Proceedings, *Welding and Cutting*, Deutscher Verband für Schweisstechnik, 167, 1980

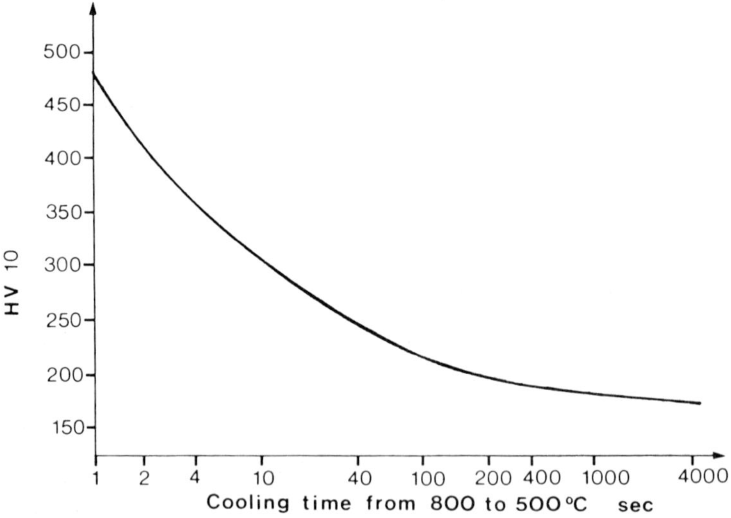

Figure 3.31 Mean Vickers hardness measurements in the HAZ of a medium strength 20 mm thick C–Mn steel based on several different welding processes. Adapted from Inagaki, M. and Sekiguchi, H., Continuous cooling transformation diagrams of steels for welding and their applications, *Transactions of the National Research Institute for Metals (Japan)*, **2,** 40, 1960

profile. Alternatively, the maximum hardness in the grain growth zone can be plotted as a function of different Δt_{8-5}, or weld heat inputs, as shown in *Figure 3.31*. This figure indicates a certain proportionality between hardness and heat input for a given steel. Note that different welding processes were used to compile this figure.

Weld CCT diagrams

In contrast to conventional CCT diagrams, weld CCT diagrams are based on a much higher austenitizing temperature, usually 1350—1400 °C, which corresponds to the grain growth zone. The curves are plotted using a weld simulator to produce the appropriate thermal cycles. An example of such a diagram is shown in *Figure 3.32* (*a*). An alternative, more comprehensive

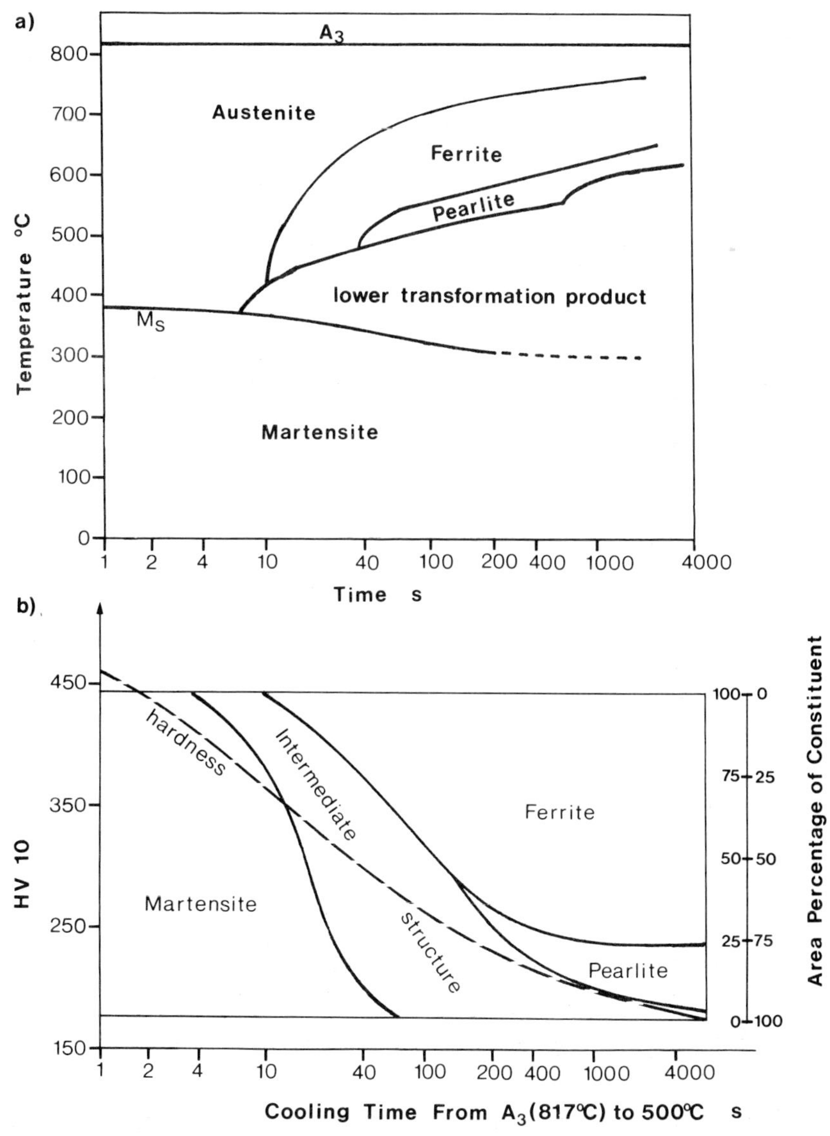

Figure 3.32 (a) Example of a weld CCT diagram for a medium strength C–Mn steel, using an austenitizing temperature of 1400 °C. (b) A microstructure–hardness diagram for the same steel as in (a). Adapted from Inagaki, M. and Sekiguchi, H., Continuous cooling transformation diagrams of steels for welding and their applications, *Transactions of the National Research Institute for Metals (Japan)*, **2,** 40, 1960

approach is one in which this diagram and *Figure 3.31* are combined, as shown in *Figure 3.32* (*b*). Clearly, such diagrams require considerable time and resources to compile and since they are specific to only one steel they have a somewhat restricted use.

Weld microstructure diagrams

Weld-microstructure diagrams have been developed by Alberry and Jones[24] specifically to help select the preheat temperature when welding creep-resistant alloys based on 0.5 Cr–Mo–V steel. An example of this diagram is shown in *Figure 3.33* from which, e.g., by knowing the type of weld process and the plate thickness the required preheat can be read off, together with the expected microstructural constituent. Thus, if a heat input of, e.g.,

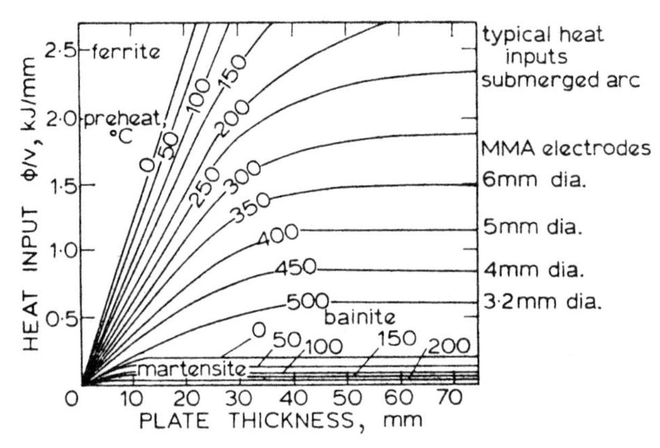

Figure 3.33 Welding microstructure diagram for a simulated fine grained 0.5 Cr–Mo–V steel. The curves denote the preheat temperature needed to obtain at least 5% ferrite, given the input energy, electrode size and plate thickness. After Alberry and Jones[24]

1.5 kJ mm^{-1} (which corresponds to a 6 mm diameter electrode in a MMA process) is used a preheat of *ca.* 350 °C is sufficient to ensure a fine grained, tough bainitic–ferritic microstructure. These diagrams are developed by using the Rosenthal and diffusion equations for estimating grain growth in the HAZ. Microstructural variations are obtained by using weld simulation to develop appropriate CCT diagrams.

Grain growth diagrams

Examples of these diagrams have already been given (*see* pp. 126–132) The advantage of grain growth diagrams over weld CCT diagrams and weld microstructure diagrams is that they require less experimental effort since they are largely based on a computing model. Thus, although grain growth in real welds or weld-simulated samples have to be made in order to check the correct activation energy for grain growth, the positions of the carbide solubility lines and grain size contours are given directly by the computer. Additional information concerning weld bead and HAZ widths can easily be incorporated for, e.g., checking maximum crack lengths in the grain growth zone and an example of this for a Nb-microalloyed steel is illustrated in *Figure 3.34*. This diagram is a plot of welding energies (here over a range typical of MIG welding) as a function of distance from the weld centre line. The fusion line, the range of

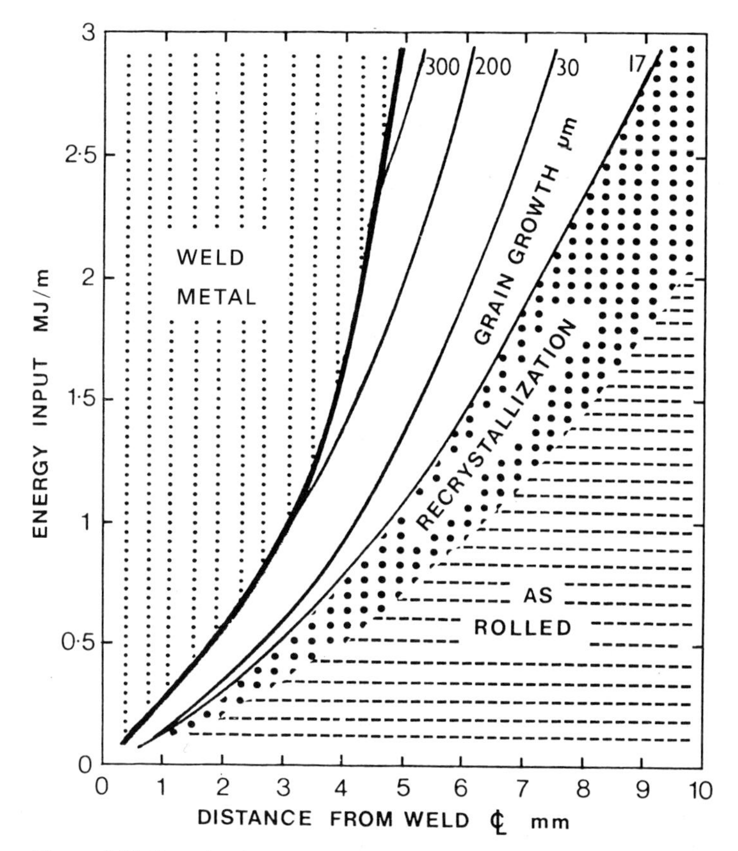

Figure 3.34 A grain size – microstructure diagram for the welding of a Nb-microalloyed steel (thick plate condition). Data for plotting the diagram is assembled mainly from the theoretical models described in the text, but is checked from real bead–on–plate welds. The range of welding input energies in this case covers MIG welding. The figure is by courtesy of J. Ion, University of Luleå

recrystallized base metal and the grain size contours are based on data from both real welds and the theoretical model. These diagrams can even be developed further by employing weld simulation to plot, e.g., the volume fraction of martensite using a point counting technique. An example of this for the case of a Nb-microalloyed steel (thick plate condition) is shown in *Figure 4.52*. As may be expected, the amount of martensite increases with increase in grain size. These diagrams can thus be used to estimate the amount of grain growth and width of the grain growth zone, the amount of martensite to be expected, and the influence of different welding processes, including preheat.

Some useful guidelines on the selection of steels to give optimum weldability and toughness have been compiled by Dolby[25], and these are summarized below.

C–Mn and microalloyed C–Mn steels with C>0.1 wt %

1. Use steels with low levels of C and low carbon equivalent if possible.
2. Use Al-treated steels to restrict the width of the grain-coarsened HAZ and to minimize the free N level (AlN is a relatively stable precipitate).
3. Use steels having a high parent steel toughness.
4. For high heat input welding, note that, for the same base C–Mn compositions, Nb and V additions lower HAZ toughness because of precipitation during cooling and reheating.
5. Use clean steels. If Al-treated or vacuum degassed this means low S levels or the use of MnS shape-control additions to improve the *z*-properties and to reduce liquation cracking problems.

Microalloyed steels with C<0.1 wt %

1. Low levels of C should produce the highest toughness, but the relevance of C_{equiv} is not clear in this case.
2. Controls 2, 3, 4 and 5 above also apply.

Low alloy steels

1. Use steels with low C-levels. A higher resistance to cleavage is obtained if the hardenability is sufficient to develop low-carbon martensite for the welding conditions used.
2. Caution should be used with post-weld heat treatment of low alloy steels that contain significant amounts of precipitation hardening elements, e.g. V and Nb and also with those that contain high levels of the residual elements Sb, P, As and Sn. Note that in Cr–Mo steels, P is probably the most important element as regards minimizing temper embrittlement, whereas in Ni-containing steels, Sb and P are probably controlling.
3. Controls 2, 3, 4 and 5 above also apply.

Multi-run welds

Refinements in microstructure, improvements in toughness and reductions in residual stress can all result from multi-run welding as compared to a single weld run of the same cross-section. The reasons for this are as follows.

1. That each subsequent weld thermal cycle effectively grain refines or 'normalizes' part of the previous weld metal

2. The total input energy per weld run is decreased so that the amount of grain growth is accordingly reduced.
3. Previous weld runs can provide a certain preheat which tends to extend the Δt_{8-5} cooling time.
4. Subsequent weld runs tend to anneal out residual stresses caused by previous runs.

The effect of a second weld run is illustrated schematically in *Figure 3.35*. It is seen that subsequent runs cause parts of earlier welds, i.e. weld metal and HAZ, to be reheated. The possible effects of this additional weld thermal cycle on the prior weld metal and its HAZ are now considered briefly.

The weld metal

On the basis of previous considerations (*see* Chapter 2), the as-solidified weld metal is likely to be fairly low alloyed, to consist of various transformation products (depending on input energy) and to contain a high dislocation density. During reheating, therefore, there is likely to be little to hinder recrystallization and grain growth, particularly in the absence of stable carbonitrides. However, since the energy input per weld run is relatively low grain growth will not be extensive.

The HAZ

The net effect of reheating the HAZ is to refine the microstructure, at least in steels, rather than to bring about larger prior-austenitic grain sizes. As can be deduced from *Figure 3.35*, only a small part of the reheated HAZ is subjected

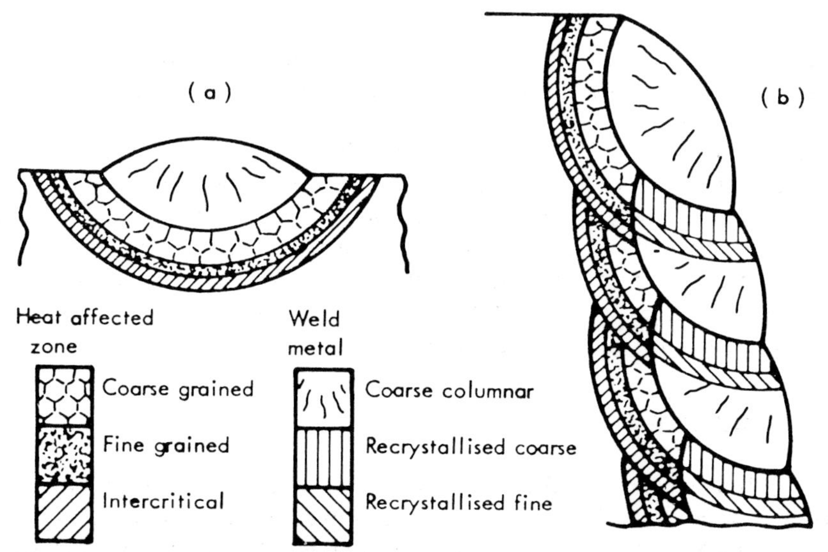

Figure 3.35 Schematic comparison of the microstructures of (a) single run and (b) multi-run welds. After Coleman, M.C., The structure of weldments and its relevance to high temperature failure, *Weldments: Physical Metallurgy and Failure Phenomena, Proceedings of the 5th Bolton Landing Conference*, p. 409, eds Cristoffel, R.J., Nippes, E.F. and Solomon, H.D., General Electric Company, 1978

to the highest peak temperatures during both weld runs. Thus, the main part o all the shaded regions experiences relatively mild heat treatments.

Another interesting feature of multi-run welds is that the larger th number of beads the greater is the volume fraction of reheated weld metal tha is produced. On the other hand, the use of fewer (larger) runs (provided ther are several weld runs) also tends to increase the amount of recrystallized wel

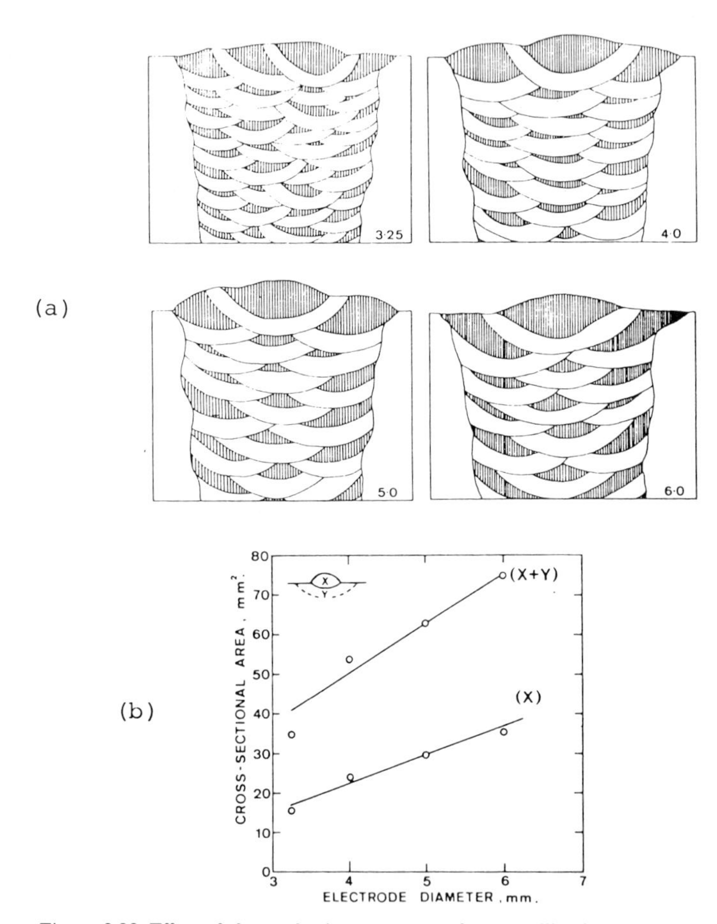

Figure 3.36 Effect of electrode size on amount of recrystallized weld metal in multi-run welding. (a) The cross-sections as a function of electrode size; the white areas represent recrystallized weld metal. (b) The graphical representation. After Evans, G.M., Factors affecting the microstructures and properties of C–Mn all-weld metal deposits, *Welding Reviews*, **4,** May 1982

metal. Both of these effects are illustrated in *Figure 3.36,* which shows (*a* cross-sections of multi-run welds of different electrode sizes, and (*b*) a plot o electrode size *versus* amount of recrystallized area. Note the slight divergenc in the curves, which denotes a certain increase in recrystallized weld meta when employing larger electrodes. *Figure 3.36* can be compared with *Figur 1.13,* which shows the cross-section of an actual multi-run weld.

The increased volume of grain refined weld metal and the possible removal of segregation, such as columnar grain-boundary carbides, result in higher notch toughnesses, *Figure 3.37*. This figure also emphasizes the difference between horizontal and vertical-up welding. Indeed, the beneficial effect of multi-layer welding is particularly demonstrated in basic weld metals that have a favourable slag system.

It would be possible, in principle, to apply the heat-flow equations of weld thermal cycling to calculate the modifications to microstructure that result from multi-run welding (for one attempt at this, *see*, e.g., ref. 26). However, in

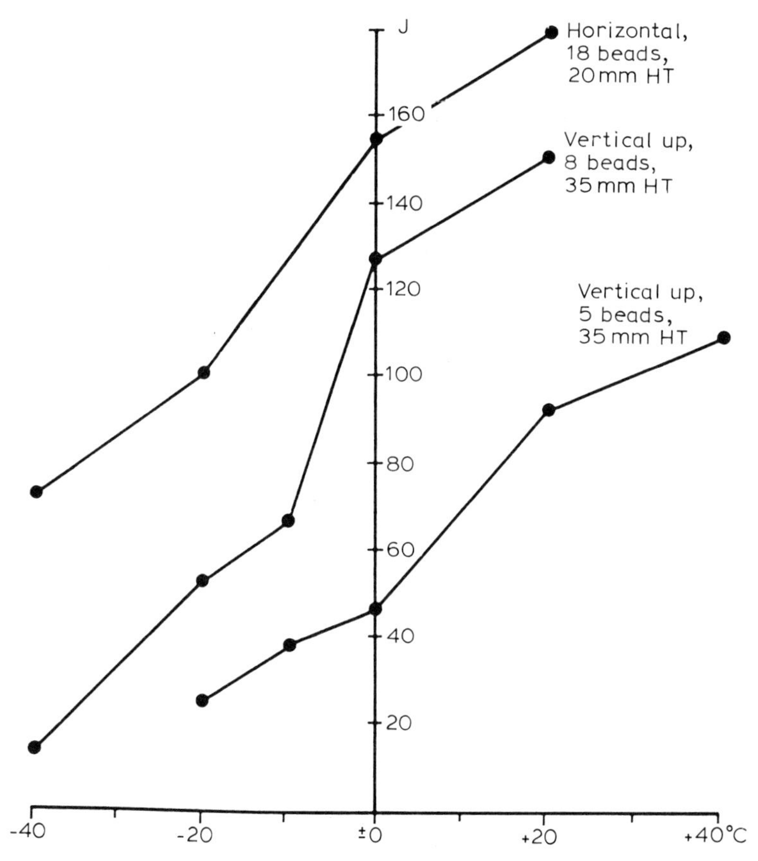

Figure 3.37 Influence of multi-run welding and welding method on transition temperatures of steel weld metals. After Almqvist, G., *et al.* (*see* ref. 5 in Chapter 1)

general, the plots resulting from single weld runs probably represent the most severe case with respect to grain growth in the HAZ, and for reasons discussed above, the additional thermal cycles that result from multi-run welding generally tend to refine the microstructure and improve the properties of the weld.

The ultimate test of predictions of weldability is how the weld behaves in practice. In many respects, the treatment given in the first three chapters of this book provides the basis for the subject of Chapter 4: what are the conditions that give rise to cracking and fracture in welds?

References

1. Morrison, W.B. and Chapman, J.A., Controlled rolling, *Rosenhain Centenary Conference Proceedings,* p. 193, eds, Baker, R.G. and Kelly, A., Royal Society, 1975

2. Gladman, T., Dulieu, D. and McIvor, I.D., Structure – property relationships in microalloyed steels, *Microalloying 75, International Symposium on High Strength Low Alloy Steels,* p. 25, Union Carbide Corp., Washington, 1975

3. Cottrell, A.H., *The Mechanical Properties of Matter,* John Wiley, 1964

4. Honeycombe, R.W.K., *Steels – Microstructure and Properties,* Edward Arnold, 1981

5. Mercer, W.L., Material requirements for pipe line construction, *Rosenhain Centenary Conference Proceedings,* p. 7, eds Baker, R.G. and Kelly, A., Royal Society, 1975

6. *Guide to the Welding and Weldability of C–Mn Steels and C–Mn Microalloyed Steels,* International Institute of Welding, Publications Document II S/IIW–382–71, 1971

7. Cotton, H.C., Material requirements for off-shore structures. *Rosenhain Centenary Conference Proceedings,* p. 19, eds Baker, R.G. and Kelly, A., Royal Society, 1975

8. Reed-Hill, R.E., *Physical Metallurgy Principles,* Van Nostrand Reinhold, 1964

9. Phillips, A., *Welding Handbook,* McMillan, 1966

10. Räsänen, E. and Tenkula, J., Phase changes in the welded joints of constructional steels, *Scandinavian Journal of Metals,* **1,** 75, 1972

11. Ashby, M.F. and Easterling, K.E., A first report on diagrams for grain growth in welds, *Acta Metallurgica,* **30,** 1969, 1982

12. Haasen, P., *Physical Metallurgy,* p. 207, Cambridge University Press, 1978

13. Ikawa, H., Shin, S., Nakao, Y., Nishimoto, K., Tanaka, T. and Nakahama, S., Study on behaviour of carbides and knife line attack phenomenon in stabilized austenitic stainless steels (report V), *Journal of Japanese Welding Society,* **46,** 402, 1977

14. Zener, C., Personal communication to Smith, C.S., *see American Institute of Mining and Metallurgical Engineers,* **175,** 15, 1949

15. Gladman, T., On the theory of the effect of precipitate particles on grain growth in metals, *Proceedings of the Royal Society,* **294,** 298, 1966

16. Alberry, P.J., Chew, B. and Jones, W.K.C., Prior austenite grain growth in the heat affected zone of a 0.5 Cr–Mo–V steel, *Metals Technology,* **4,** 317, 1977

17. Ikawa, H., Oshige, H. and Noi, S., Study on the grain growth in weld heat affected zone (report 5), *Journal of Welding (Japan)*, **7,** 396, 1977

18. Strid, J. and Ion, J., University of Luleå, unpublished research, 1981

19. George, T. and Irani, J.J., Control of austenitic grain size by additions of titanium, *Journal of the Australian Institute of Metals,* **13,** 94, 1968

20. Matsuda, S. and Okumura, N., Effect of distribution of TiN precipitate particles on the austenite grain size of low carbon low alloy steels, *Transactions I.S.I.J.,* **18,** 198, 1978

21. Kanazawa, S., Effect of distribution of TiN precipitate particles on the austenite grain size of low carbon low alloy steels, *Transactions I.S.I.J.,* **16,** 486, 1976

22. Hannerz, N-E., Swedish Board of Technical Development (STU), Report No 70–728/U 615, Stockholm 1972

23. Houghton, D.H., Weatherley, G.C. and Embury, J.D., *Characterization of Carbonitrides in Ti-bearing HSLA Steels,* McMaster University Report, 1981

24. Alberry, P.J. and Jones, W.K.C., Diagram of the prediction of weld heat-affected zone microstructure, *Metals Technology,* **4,** 360, 1977

25. Dolby, R.E., *Factors Controlling Weld Toughness – the Present Position,* The Welding Institute, Report No 14/1976/M, Abington, May 1976

26. Alberry, P.J. and Jones, W.K.C., A computer model for the prediction of heat affected zone microstructures in multipass weldments, *Metals Technology,* **9,** 419, 1982

Further reading

Castro, R.J. and de Cadenet, J.J., *Welding Metallurgy of Stainless and Heat Resisting Steels,* Cambridge University Press, 1974

Christian, J.W., *Theory of Transformations in Metals and Alloys,* Pergamon Press, 1965

Dolby, R.E., *Factors Controlling Weld Toughness — the Present Position,* The Welding Institute, Report No 14/1976/M, Abington, May 1976

Guide to the Welding and Weldability of C–Mn Steels and C–Mn Microalloyed Steels, International Institute of Welding, Publications Document IIS/IIW-382-71, 1971

Honeycombe, R.W.K., *Steels — Microstructure and Properties,* Edward Arnold, 1981

Hrivňák, I., *'The Theory of Mild and Micro-alloy Steels' Weldability,* The Welding Institute, Bratislava, 1969

Porter D.A. and Easterling, K.E., *Phase Transformations in Metals and Alloys,* Van Nostrand Reinhold, 1981

Chapter 4

Cracking and fracture in welds

The severe thermal cycle and the high restraints involved in welding thick, strong metals make it inevitable that cracking problems sometimes occur. Cracks in welds can broadly be divided into two main categories:

1. Those attributed to the welding process itself.
2. Those occurring during service.

The latter is mainly concerned with external factors that can degrade the weld and cause cracking, such as the effects of environment, vibration or thermal cycling. In this chapter only the former category is considered, i.e. those that arise from the process of welding itself, including reheat annealing.

The types of cracking phenomena associated with the welding process include:

1. Hydrogen attack (weld deposit).
2. Gas porosity (weld deposit).
3. Solidification cracking (weld deposit).
4. Liquation cracking, HAZ burning or hot tearing (fusion zone).
5. Lamellar tearing (HAZ).
6. Cold cracking or hydrogen cracking (HAZ).
7. Reheat cracking (weld deposit and HAZ).

Of these, the most interesting with respect to complexity and severity are usually considered to be solidification cracking, liquation cracking, lamellar tearing, cold cracking and reheat cracking, and these phenomena are the main concern herein. As a basis for discussion of the mechanisms of cracking in the various cases, it is first useful to consider the question of what constitutes the crack resistance, or *fracture toughness,* of a material and how welding can affect this property.

Fracture toughness

It is usual in engineering to talk of *ductile* or *brittle* fracture. A few of the factors that govern the ductile–brittle cleavage transition are discussed in Chapter 3 (pp. 105–112). Examples of the fracture surfaces from ductile and

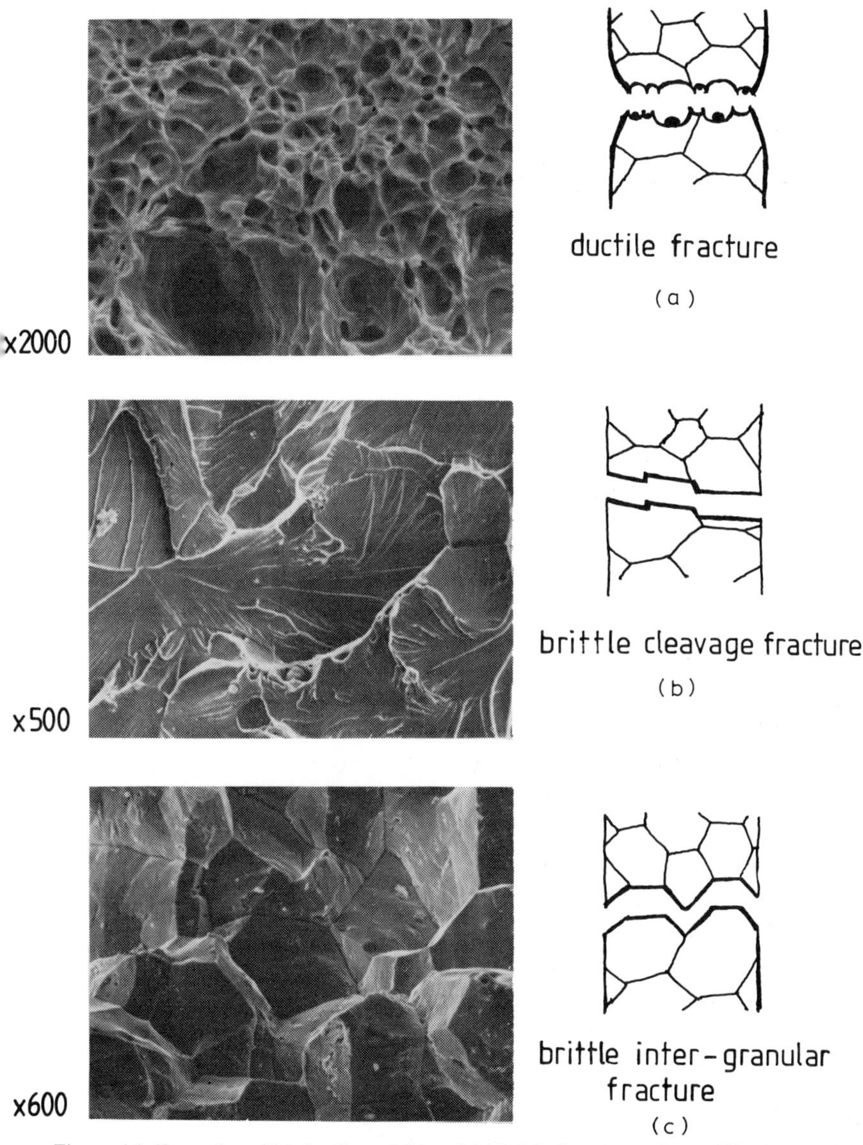

Figure 4.1 Examples of (a) ductile and (b) and (c) brittle fracture surfaces. All micrographs are taken in the scanning electron microscope. Courtesy of R. Harrysson, University of Luleå (reduced by two-fifths in printing)

brittle failures in steels are shown in *Figure 4.1*. The ductile failure is characterized by 'dimpling' of the surface due to localized plastic flow occurring around small inclusions, *Figure 4.1* (*a*). The brittle cleavage fracture surface, on the other hand, appears comparatively flat and featureless and exhibits little or no evidence of plasticity, *Figure 4.1* (*b*). In an optical stereomicroscope, the brittle cleavage fracture is easily recognized because of the brightly reflecting characteristics of the freshly broken transcrystalline (usually $\{100\}_{\alpha}$) surface facets. If, on the other hand, fracture is intercrystalline, i.e. the fracture path has followed along grain boundaries, the fracture surface is less bright and the grain shapes are recognizable, *Figure 4.1* (*c*). As a general rule, high fracture toughness is associated with the amount of energy it takes for a crack to propagate through the material, and this in turn is roughly proportional to its path length and the energy of plastically deformed material (compare, e.g. *Figures 4.1* (*a*) and (*b*)). Fracture toughness is thus mostly concerned with distinguishing between the conditions that characterize high and low crack resistances in materials, and thereby with defining the exact condition under which brittle fracture occurs. For an excellent introduction to this subject, as applied to welding, *see* ref. 1. Although the existence of a crack or sharp notch is a necessary prerequisite for brittle failure, its mere presence is no guarantee that brittle failure results, as is shown later.

Consider a notched sample under a uniaxial tensile stress, σ_{app}, as shown in *Figure 4.2* (*a*). The notch or crack induces a stress intensity profile with the

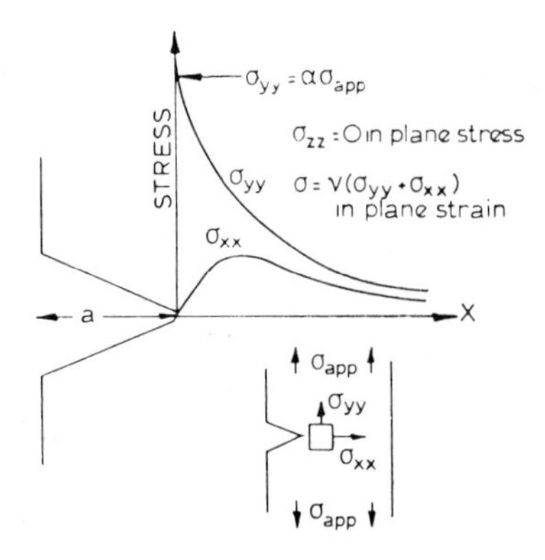

a) stresses at a notch

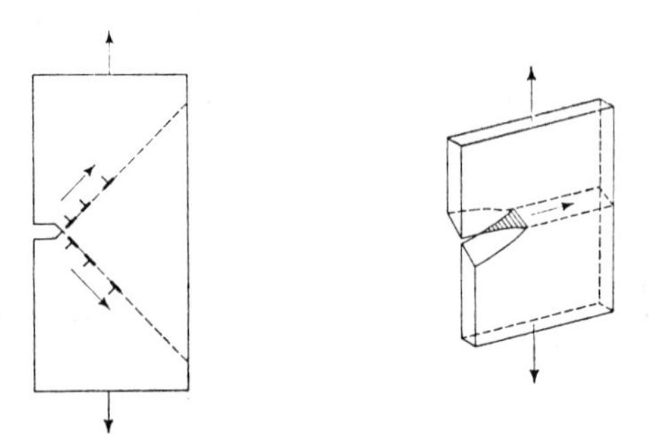

b) plane strain c) plane stress

Figure 4.2 (a) The elastic stress distribution in a notched specimen. After Knott, J.F. Second phase particles and the toughness of structural steels, *Effect of Second-phase Particles on the Mechanical Properties of Steel,* Iron and Steel Institute, London, 1971. (b) and (c) Comparison between plane-strain and plain-stress fracture in ductile metals. After Cottrell, A.H., *Mechanical Properties of Matter,* John Wiley, 1964

principal stresses σ_{xx}, σ_{yy} and σ_{zz}. Under what conditions can the crack grow in length? If the metal is ductile, the region at the crack tip deforms plastically under the influence of the high stress intensity (σ_{yy}), and this causes the crack faces to move apart. The more ductile the metal, the larger the plastic zone at the crack tip becomes. Suppose now that the material is sufficiently hard that growth of the plastic zone is restrained by the surrounding material. This restraint causes secondary elastic stresses to build up around the crack tip, both at right angles to the applied stress in the plane of the crack (σ_{zz}), and also parallel to the crack face (σ_{xx}). Thus, a triaxial state of tensile stress is developed. The state of triaxial stress which allows no contraction or strain in the through-thickness direction is defined as a condition of *plane strain, Figure 4.2* (*b*). If there are no restraints developed in the thickness direction the condition is defined, instead, to be one of *plane stress, Figure 4.2* (*c*). Thus,

$$\sigma_{zz} = 0 \tag{4.1}$$

equation (4.1) can be applied to the case of plane stress and equation (4.2) to

$$\sigma_{zz} = \nu\ (\sigma_{yy} + \sigma_{xx}) \tag{4.2}$$

that of plane strain, where ν is Poisson's ratio. On this basis it can be concluded that, if the sample is thick enough, a notch can bring about a triaxial state of stress which locally increases the yield stress of the material ahead of the crack. Since the conditions that give rise to restraint increase if the thickness (xz) of the material is larger or if the plastic zone is smaller, it follows that a triaxial stress is more likely to occur in thicker sections or harder material. The possible ways of hardening the material are by decreasing the temperature, increasing the rate of loading or changing the microstructure.

If the maximum triaxial stress at the crack tip exceeds the fracture stress, σ_f, of the material the crack can grow. As shown by equation (3.6), once the crack size c increases, σ_f is effectively lowered and crack growth can even accelerate. It can be shown that the stress distribution over a distance, x, ahead of a notch is given by equation (4.3), where K is a *stress intensity factor* that is

$$\sigma_{yy} = \frac{K}{(2\pi x)^{1/2}} \tag{4.3}$$

dependent upon the applied stress, *see* equation (4.4) where $2c$ is the diameter

$$K = \sigma_{\mathrm{app}}\ (\pi c)^{1/2} \tag{4.4}$$

of the thin disc-shaped crack. When failure occurs at some critical value of the applied stress, σ_{app}, the parameter K assumes a value defined as K_c. K_c is considered to be a characteristic of the material and is termed *fracture toughness*. Strictly speaking, K_c should only be used in connection with truly brittle materials. If, however, sufficient yielding occurs to invalidate a purely elastic approach, then a modified crack length, $(c+r_y)$ can be used in equation (4.4), r_y being the radius of the plastic zone.

It is important to relate K_c to the triaxial stress state of the material. As pointed out above, plane strain is achieved when the material has a certain minimum thickness (or hardness), and the corresponding toughness of the material is then termed *plane strain fracture toughness*, K_{IC}. This is often taken as a constant for a material. For example, K_{IC} for structural steels is *ca.* 100 N mm$^{-3/2}$ and steels with values below *ca.* 10 N mm$^{-3/2}$ are considered to be brittle. However, in practice, K_{IC} is dependent upon changes in strain rate and temperature. This interdependence between thickness, strain rate and temperature is sometimes neglected in fracture toughness testing where high strain-rate tests, such as Charpy, are used to estimate the fracture toughness of thick plates that may be subjected to relatively low rates of strain in service. It is useful to consider tests used for estimating fracture toughness in more detail.

Fracture toughness testing

The most direct technique for measuring the fracture toughness is called the *crack opening displacement* (COD) test. It is based on the distance, δ, that a crack can open up under stress without the crack increasing in length. It can be shown that δ is related to K_c by equation (4.5), where σ_y is the yield stress and ε_y

$$\left(\frac{K_c}{\sigma_y}\right)^2 = \frac{\delta_c}{\varepsilon_y} \tag{4.5}$$

is the yield strain of the material. The useful feature of COD is that the fracture toughness of ductile materials can be measured directly using, e.g. a simple clip gauge. COD measurements that involve different thicknesses of test sample are shown in *Figure 4.3*, which illustrates very well how dependent the tran-

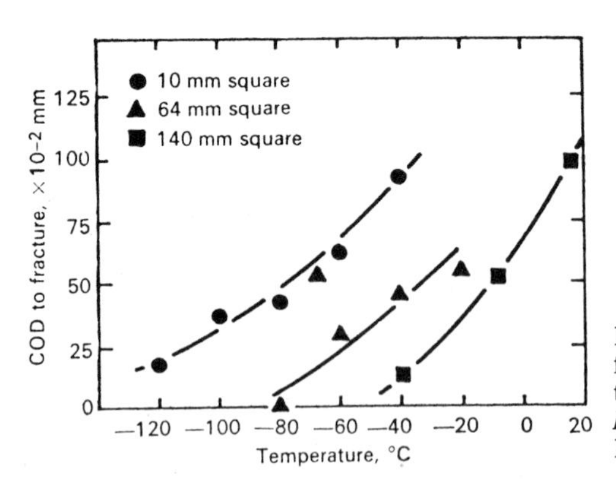

Figure 4.3 Crack opening displacement as a function of temperature in different sizes of test samples. After Richards, K.G., *Brittle Fracture of Welded Structures*, The Welding Institute, 1971

sition temperature is on specimen size. *Figure 4.4* shows how the fracture toughness of the parent metal compares with that of the various zones of the HAZ and weld metal. The effect on transition temperature of strain ageing in the heat-affected base metal is clearly distinguished in *Figure 4.5*.

It is useful to estimate a critical size of crack for a given material, loading condition and operating temperature which marks the onset of rapid fracture.

From equation (4.4) fracture occurs when $c = c_{crit}$, *see* equation (4.6), where C

$$c_{crit} \approx C \left(\frac{K_{IC}}{\sigma_y}\right)^2 = C\,\frac{\delta_{IC}}{\varepsilon_y} \quad \text{mm} \tag{4.6}$$

is a constant dependent on the applied stress conditions (*see*, e.g., ref. 2). Note that an increase in hardness of the material effectively decreases the critical size of crack.

A disadvantage of COD testing is the fact that the test samples must be pre-cracked. The usual technique is to produce a fatigue crack at the root of a

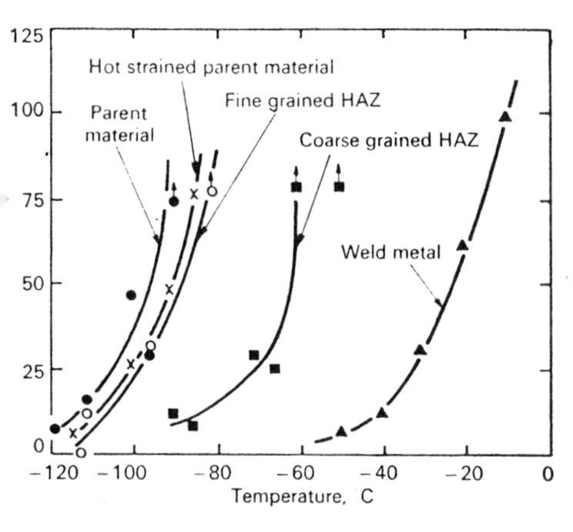

Figure 4.4 Comparisons of crack opening displacement between different zones in the HAZ. After Richards, K.G., *Brittle Fracture in Welded Structures*, The Welding Institute, 1971

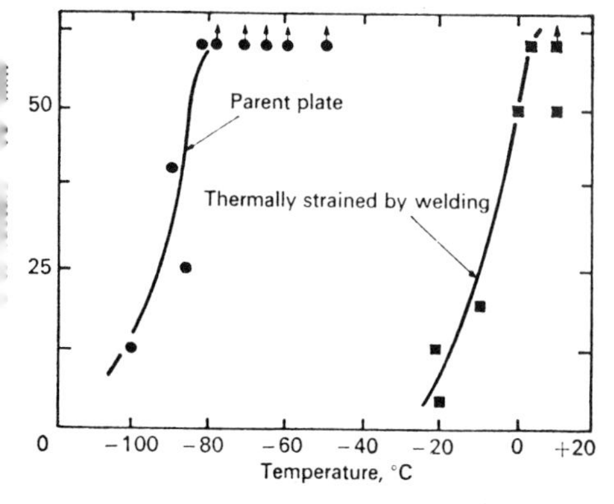

Figure 4.5 The effect of strain ageing due to welding on crack opening displacement. After Richards, K.G., *Brittle Fracture in Welded Structures*, The Welding Institute, 1971

machined notch (*see Figure 4.6* (*a*)), a process which is generally time consuming and expensive. Furthermore, there can be difficulties in obtaining consistent values of δ_c for a given material, and considerable experimental scatter may result.

The most widely used technique to evaluate the impact properties or toughness of welds is the *Charpy V-notch test*. Typical dimensions and layout of this test are shown in *Figure 4.6* (*b*). The purpose of the test is to plot the energy absorbed in fracturing a specimen using a pendulum hammer, and hence to relate the energy absorbed due to the impact with the specimen toughness. A

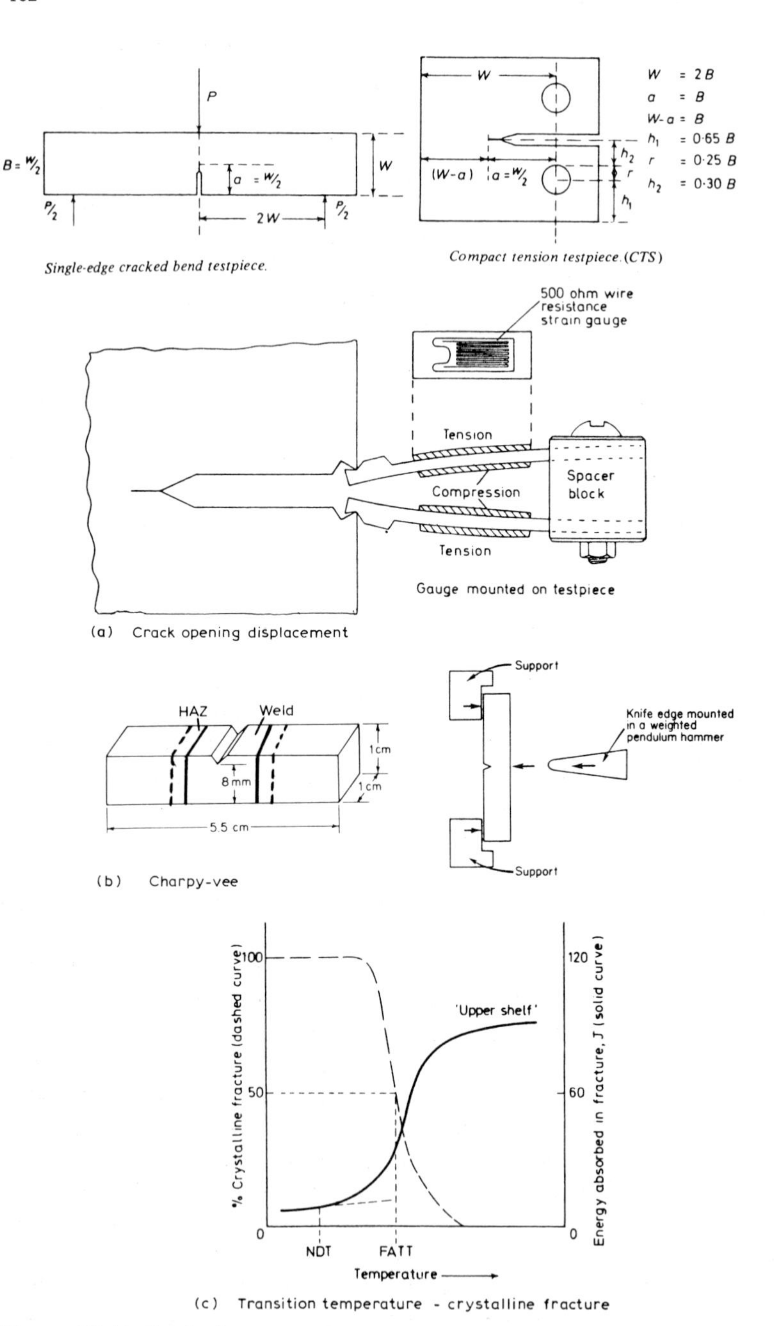

Figure 4.6 (a) Standard test samples and method of measuring crack opening displacement. (b) Charpy V-notch test sample. (c) Schematic impact transition curves for steel. (a) and (c) are after Knott, J.F., *Fundamentals of Fracture Mechanics*, Butterworths, 1973. (b) is after Reed-Hill, R.E., *Physical Metallurgy Principles*, Van Nostrand Reinhold, 1964

typical mean plot of results for a structural steel is illustrated in *Figure 4.6* (*c*), in which the energy absorbed and the percentage of crystalline (cleavage) fracture are given as a function of testing temperature. It is seen that there is a transition from low- to high-energy fracture over a relatively narrow temperature range and that this is associated with a change from transcrystalline to fibrous, or dimpled, ductile fracture. Thus, material quality can be defined in terms of this *transition temperature,* which in the welding literature is usually specified as a certain minimum energy (*ca.* 27 J), which has to be absorbed at a given test temperature. Alternatively, the transition temperature is often simply defined in terms of, e.g., 50% crystalline fracture as estimated from the broken Charpy specimen. In practice, as with COD testing, there is considerable scatter in experimental results for a given material when these curves are plotted.

In welding applications the notch is typically located at the weld centre line, as illustrated in *Figure 4.6* (*b*). For HAZ testing, notch location is more difficult and it is usually better to use weld simulated samples instead.

The Charpy test generally reproduces the ductile–brittle transition temperature in steel at about the correct temperature in relation to the bulk structure[3]. It is also thought to rank material toughness in the right order and this is particularly useful in quality control testing. A further important consideration is that test bars are cheap to produce and the test is a simple one to perform. On the negative side, it has been argued that the notch is too blunt to be representative of real cracks in a material. Furthermore, the high impact velocity of the pendulum hammer, giving strain rates at the notch of *ca.* 10^5–10^6 s^{-1} (i.e. about a million times faster than conventional tensile tests!), cannot usefully be compared with strain rates likely to be experienced by welds in normal service. Yet another criticism concerns the standard specimen size of 1 cm^2, which may give misleading results when testing for much thicker weld geometries.

To summarize, since by far the greatest tonnage of metals used today in welded construction belong to ductile or 'yielding materials', the most realistic approach to brittle fracture evaluation is that based on the so-called *general yielding fracture mechanics.* Crack opening displacement provides a useful method of determining the critical flaw size, and the method is very sensitive to changes in sample thickness and hardness or strength. The Charpy test, widely used in the welding industry, provides a useful quality control test for ranking the toughness of materials, and provides a fairly reliable value of the transition temperature. However, because of the high strain rates employed in this test, the measured toughness may tend to be very pessimistic in relation to welded constructions not normally loaded in this way. On the other hand, the Charpy test specimen size may be less than that actually used in the construction, implying that the toughness derived is too optimistic. Variables such as these make interpretation of Charpy results rather more difficult than those of COD. For further discussion of the fracture-toughness testing of welds, *see,* e.g., refs 2 and 4. How some of the fracture toughness criteria examined may help to explain the various types of cracking found in welds is discussed in the following sections.

Solidification cracking

Solidification cracking occurs in weld deposits during cooling of the weld and examples are shown in *Figures 4.7* and *4.8*. These cracks occur predominantly at the weld centre line or between columnar grains. The latter form of cracking is often referred to as 'dovetail' cracking because of the shape of the weld associated with these cracks, e.g. as in *Figure 4.8* (*a*). Solidification cracking is particularly associated with end cracking, i.e. the end of, e.g., a fillet weld of a thick plate, *Figure 4.9*. Cracking is typically observed to occur at temperatures of about 200—300 °C below T_m.

The susceptibility of weld metal to solidification cracking appears to depend on three main factors, these being:

1. The coarseness of the solidification microstructure.
2. The amount and species of segregation.
3. The geometry of the joint.

All of these factors have been discussed in some detail already (*see* Chapter 2), and are summarized below.

Solidification structure

Epitaxial solidification (*see* pp. 56–59) causes the coarseness of the weld deposit microstructure to be inherited from the grain growth zone of the HAZ. Thus, high energy welds, e.g. electroslag and submerged arc, in general give the largest grain growth and hence the coarsest microstructure in the weld deposit.

The solidification patterns are affected by the welding speed (*see* pp. 49–56). Low speeds tend to allow the growing columnar grains to follow the arc, curving in behind the moving heat source. This also has a certain grain refining effect because new grains have to nucleate in order to maintain growth along <100> directions. High speed welds, on the other hand, tend to produce solidification patterns in which the columnar crystals grow in parallel, straight rows to the weld centre line. This may even result in sudden and abrupt changes in growth direction at the centre line, as the final part of the weld that solidifies attempts to keep up with the moving arc. In general, the long, straight sided columnar grain structure tends to be weaker under stress than the more equiaxed, finer grain structure of the slow speed weld. On the other hand, the cell structure may be coarser and segregation higher for the case of low welding speeds (*see* later).

The onset of cellular patterns within the columnar grain which are due to constitutional supercooling (*see* pp. 66–73), has only a minor effect on cracking susceptibility in that it tends to reduce segregation at the high angle boundaries. Thus, the coarser the cell structure, the higher the segregation tendency. High speed welding tends to produce finer cell spacing than slow speed welding.

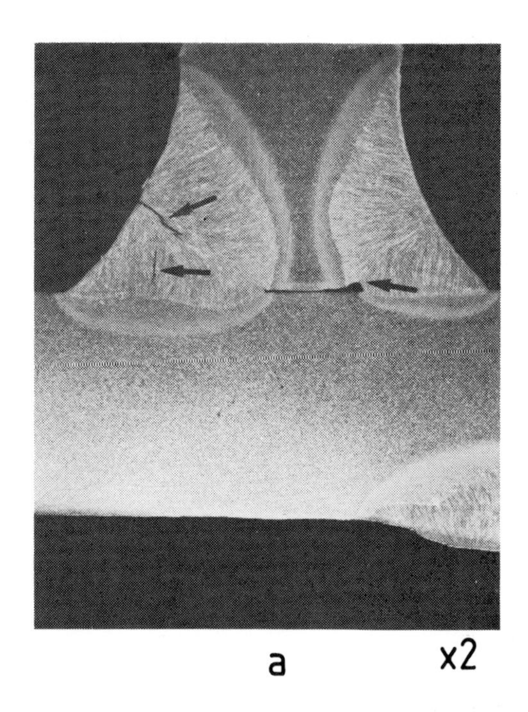

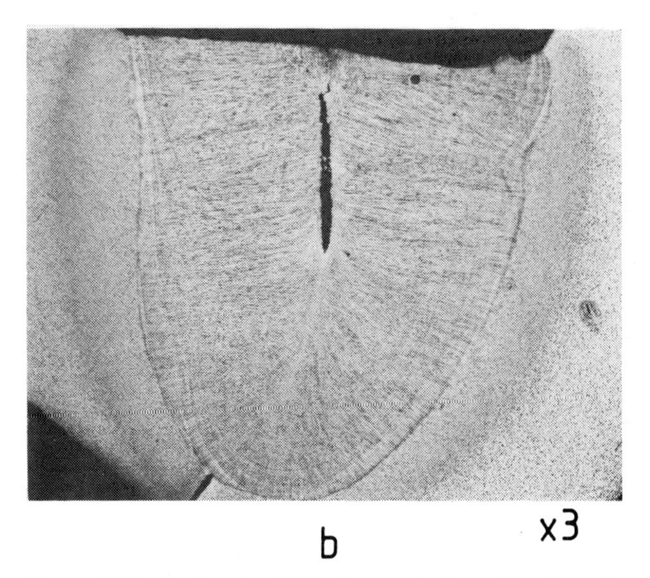

Figure 4.7 Examples of solidification cracks in welds. Several cracks are visible in (a) as arrowed. The severe crack in (b) results from a very high restraint join. Both micrographs are by courtesy of ESAB, Gothenburg (reduced by one-fifth in printing)

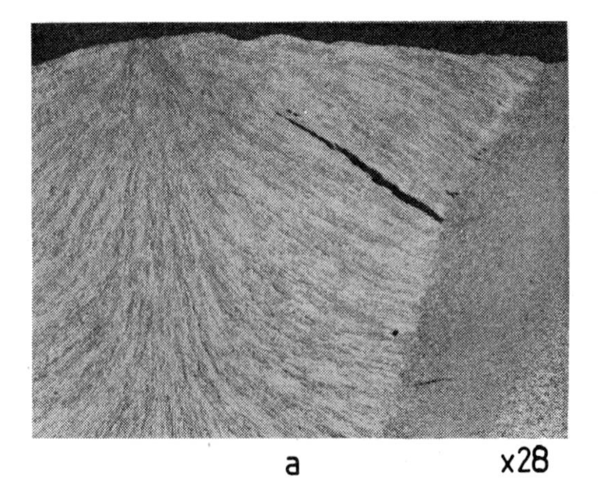

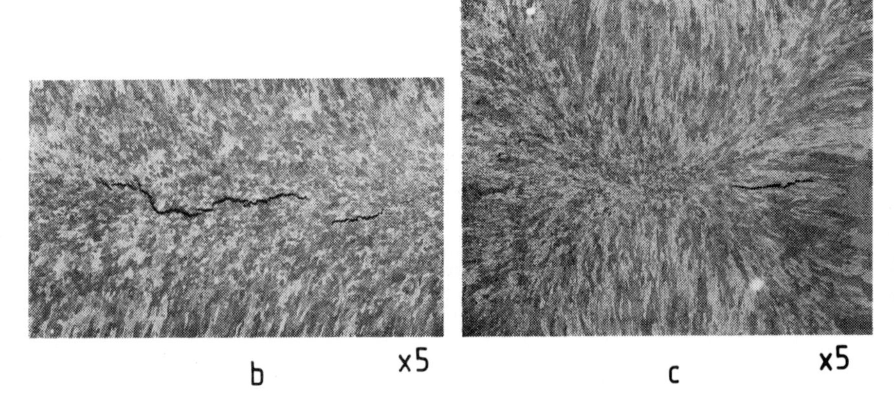

Figure 4.8 (a) Example of a solidification crack occurring at columnar grain boundaries. The micrographs (b) and (c) are different sections of the same crack in a submerged arc weld. Micrographs are by courtesy of ESAB, Gothenburg (reduced by two-fifths in printing)

The cellular to dendritic transition, associated with the final stages of solidification at the centre line of high energy welds, is likely to affect cracking susceptibility. This is because dendritic growth in weld metals is the result of high constitutional supercooling, relatively slow cooling rates and hence larger segregation. The problem is greatest at end-craters where constitutional supercooling is at its highest.

Segregation

Segregation cannot be avoided in alloy weld solidification. The amount of segregation and its effectiveness, however, can be controlled to some extent. Segregation occurs because of partitioning of elements during the initial stages of solidification (*see* pp. 59–65). The amount of segregation at this stage depends on the partitioning coefficient, k (equation (2. 11)), *see* equation

$$k = \frac{X_S}{X_L} \tag{4.7}$$

(4.7), where X_S and X_L are mole fractions of solute in the solid and liquid phases at a given temperature, as defined approximately by the phase diagram (*see*, e.g., *Figure 2.14*). Thus, different alloying combinations have quite different k values, the greatest partitioning occurring for the smallest values of k. In the case of steels, for example, any alloying constituent in the weld deposit which exhibits a wide freezing range of composition with iron is likely to have a low value of k. Some approximate values of k for alloys of iron, as determined

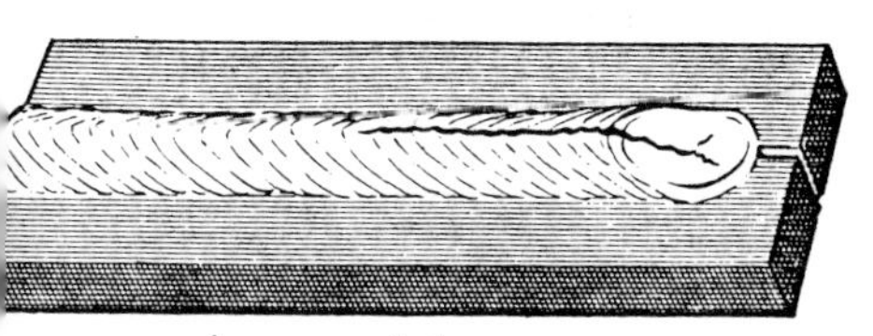

end cracking

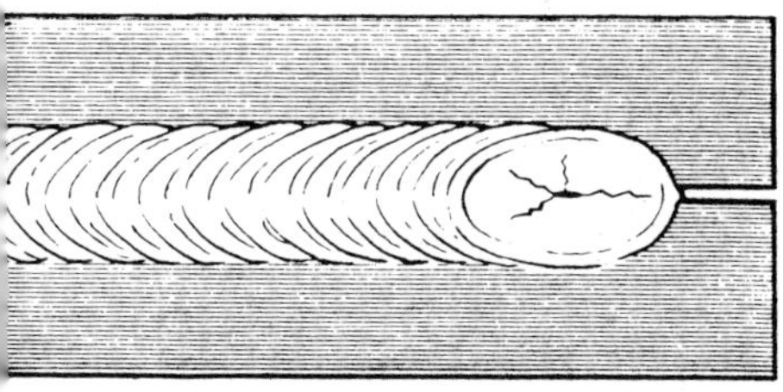

end-crater cracking

Figure 4.9 Examples of end cracking in fillet welds associated with high constraint joins. After Richards, K.G., *Weldability of Steel*, The Welding Institute, 1967

from their binary equilibrium solubilities, are given in *Table 4.1*. According to this, the elements most likely to segregate in steel are S, O, B, P, C, Ti, N, and H, in that order. Of these elements, S is often considered to be the most dangerous because it readily combines with iron and manganese to form (MnFe)S, a compound of low melting point which easily spreads along grain boundaries.

The total amount of segregation is a function of k, the atomic percent of the element present and the coarseness of the microstructure. As a general rule, slow solidifying welds with coarse microstructure, such as electroslag, tend to promote centre line segregation. This problem is discussed, for example, on pp. 59–65, *see Table 2.1* (p. 63), where it is shown that solute enrichment at the centre line of electroslag welds can extend to some mm in width. Even in manual metal arc welding, centre line segregation can be several tens of grain boundary widths in thickness. Since even monolayer grain boundary contamination from such elements as S can substantially reduce intergranular strength, it is hardly surprising that hot cracking problems may occur.

Segregation is closely associated with the phenomenon of constitutional supercooling. High segregation levels thus tend to cause high constitutional supercooling and this, in turn, brings about coarser cellular–dendritic solidification patterns. As stated earlier, high segregation levels and relatively small

TABLE 4.1 Estimations of the partitioning coefficient of elements in δ-iron (after Kumar, R., *Physical Metallurgy of Iron and Steel,* Asia Publications, 1968)

Element	Al	B	C	Cr	Co	Cu	H	Mn	Mo
k	0.92	0.05	0.13	0.95	0.90	0.56	0.32	0.84	0.80
Element	Ni	N	O	P	Si	S	Ti	W	V
k	0.80	0.28	0.02	0.13	0.66	0.02	0.14	0.95	0.90

temperature gradients across the weld give the largest amount of constitutional supercooling. In particular, the final crater at the end of a weld tends to be dendritic and has high segregation levels.

Residual stresses and joint geometry

The thermal cycle of the weld process always results in residual stresses being left in the weld. The severity of these stresses depends on the degree of restraint offered by the welded joint. As a general rule, the thicker or stronger the plates being welded, the higher the restraint and hence the greater the residual stress. At its maximum level, the tensile stresses at the weld may reach the yield strength of the metal, *see* pp. 33–45. These stresses are usually largest at the end sections of welded plates, *see*, e.g., *Figure 1.38.*

Mechanism of solidification cracking

The causes of solidification cracking are in general well understood. The partition and rejection of alloying elements at columnar grain boundaries and ahead of the advancing solid–liquid interface cause marked segregation, as illustrated in *Figure 4.10.* The segregants form low melting phases or eutectics with the metal to produce highly wetting films at grain boundaries. These films weaken the structure to the extent that cracks form at the boundaries under the influence of the tensile residual stresses that occur during cooling.

Impurities or alloying elements most likely to cause solidification cracking have the following characteristics:

1. A low partition coefficient, *k*.
2. Readily form compounds with the metal.
3. Compounds have a low melting point or form eutectics with the base metal.
4. Have a low 'wetting angle' with the metal and thus possess the ability to spread along grain boundaries.

Any alloying process which reduces the effect of segregating impurities to a minimum is obviously advantageous. For example, adding Mn where possible to form MnS instead of FeS is usually beneficial. This is because FeS has a much lower wetting angle than MnS, so that inclusions, rather than grain boundary films, are formed. It should, in fact, be noted that FeS and its eutectics freeze at lower temperatures than the manganese equivalents. FeS freezes at 1190 °C, FeS–FeO eutectics freeze at 940 °C, MnS freezes at 1600 °C and MnS–MnO eutectics freeze at *ca.* 1300 °C (from ref. 5). In some cases, higher oxygen contents in weld deposits may also be advantageous when forming oxysulphide inclusions[6]. In certain cases it may even help to substantially increase the amount of segregant. For example, in the welding of Al–Mg–Zn alloys, it is

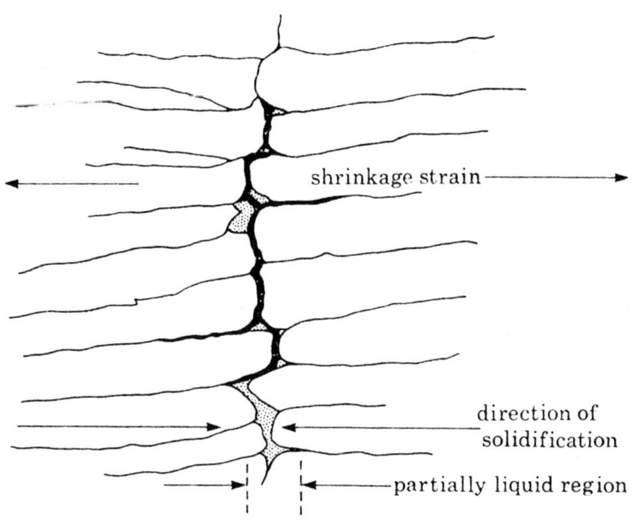

Figure 4.10 Schematic illustration of the mechanism of solidification cracking. After Baker, R.G., Weldability and its implications for materials requirements, *Rosenhain Centenary Conference Proceedings,* p. 129, eds. Baker, R.G. and Kelly, A. Royal Society, 1975

thought that Mg is the prime cause of hot cracking. However, by increasing the wire composition from 1–7% Mg, it is found[7] that hot cracking susceptibility decreases, evidently because of an increase in the amount of Al–Mg–Zn eutectic, which tends to refine the solidification structure (*see also* pp. 59–65).

The mechanism by which solidification cracking occurs is one in which cracks nucleate at carbide–matrix interfaces and spread along the boundary, under the influence of the tensile thermal stresses.

The criterion as to whether cracks nucleate within the compound, or at the compound–matrix interface, is decided by their relative fracture strengths (unless, of course, the phase is still liquid). Assuming brittle fracture occurs, the fracture strength of the compound is (from equation (3.5)) given by equation (4.8), where γ_c is the free surface energy of the compound. The

$$\sigma_f^c = \left(\frac{2E\gamma_c}{\pi c}\right)^{1/2} \tag{4.8}$$

fracture strength of the compound–matrix interface is given by equation (4.9),

$$\sigma_f^{\,i} \approx \frac{2\gamma_a}{\bar{b}} \tag{4.9}$$

where $\bar{b}$ is the width of crack at the interface and is approximately equal to one Burgers vector of matrix. γ_a is defined as the *work of adhesion* of the interface, equation (4.10), where γ_m and $\gamma_{c/m}$ refer to the matrix and interfacial surface

$$\gamma_a = \gamma_m + \gamma_c - \gamma_{c/m} \tag{4.10}$$

energies. Because of the small value of $\bar{b}$, σ_f^i is normally quite large. For example, the cementite–iron interface has a high value of γ_a and interfacial fracture rarely occurs, e.g., in pearlite. On the other hand, γ_a for MnS–Fe is close to zero[8] and experiments suggest that γ_a for FeS–Fe interfaces is also very low. Fracture is thus likely to occur primarily at the interface in this case, too.

As pointed out earlier, the role of thermal stresses is very important in solidification cracking. As a general rule, end cracking is associated with high transverse stresses (*see Figure 1.38* (*b*)), and centre line cracking behind the weld probably results from high longitudinal stresses (*Figure 1.38* (*a*)). Dovetail cracks are the result of high longitudinal stresses in association with a relatively coarse columnar grain size, as inherited from the coarse grains at the concave regions of these types of weld (*see*, e.g., *Figures 3.20* and *4.8* (*a*)). For a discussion of methods for testing susceptibility to solidification cracking *see* ref. 9 and *Table A1* (p. 223).

Liquation cracking

An example of liquation cracking is shown in *Figure 4.11*. The causes of this type of cracking are fairly well understood and are associated with grain boundary segregation aggravated by melting of boundaries near the fusion line. High residual stresses that occur as the weld cools then tend to rupture these impurity-weakened boundaries. Note that the origin of segregation in liquation cracking is quite different from that of solidification cracking. As explained on pp. 138–141, melting of grain boundaries at the fusion line occurs at temperatures between the solidus and liquidus boundaries. Since melting nucleates preferentially at high energy crystal defects, such as surfaces and boundaries, there is a gradual increase in melted boundary width up to the melt zone. Impurities of low solubility in the matrix or from melted back inclusions near the fusion line tend to diffuse to the melted boundaries. The impurities most likely to segregate under these conditions have been discussed on pp. 138–141, *see* equation (3.36) and *Figure 3.27*.

On being cooled, these segregants tend to form films of low melting point grain boundary compounds or even low melting point eutectics. An example of sulphide eutectics at the fusion zone of a welded 0.4 wt % C rail steel is shown in *Figure 4.12*. The criteria that govern liquation cracking are thus very similar to those discussed in the previous section concerning solidification cracking.

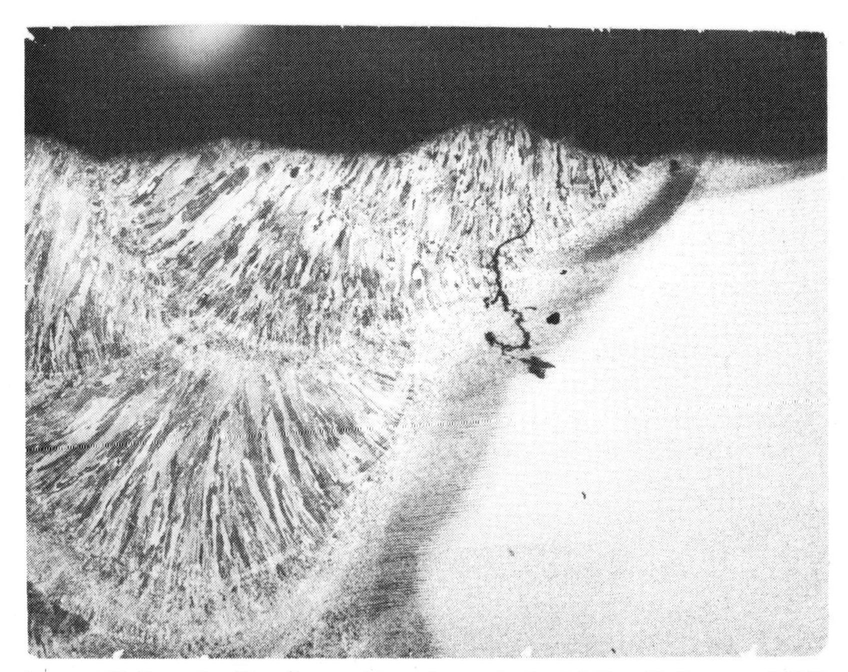

Figure 4.11 Example of liquation cracking or hot tearing in a MMA weld. Courtesy of ESAB, Gothenburg (×6)

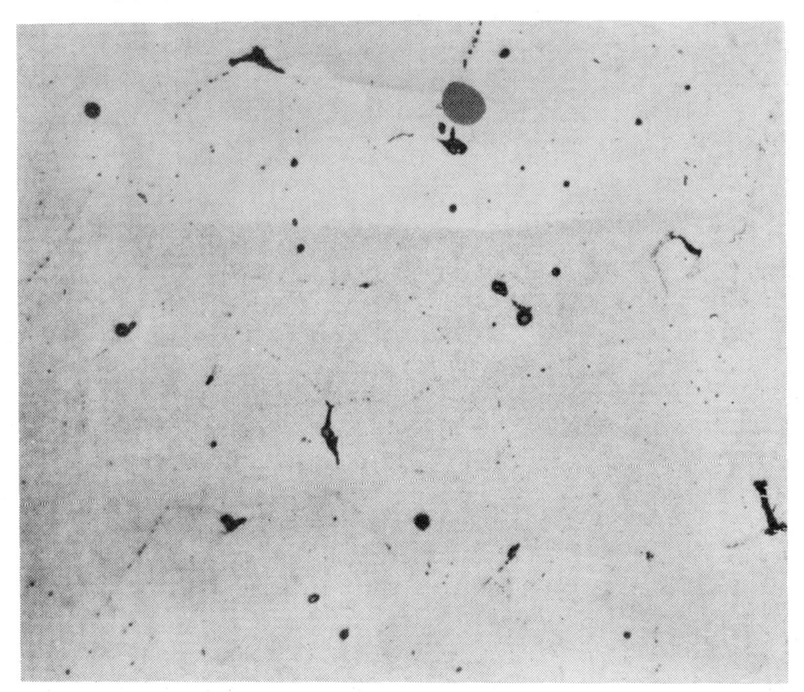

Figure 4.12 Sulphide eutectic in the fusion zone of a welded rail steel. After Baker, R.G., Weldability and its implications for materials requirements, *Rosenhain Centenary Conference Proceedings*, p. 129, eds Baker, R.G. and Kelly, A., Royal Society, 1975 (×400)

Liquation cracking is dependent on the amount and type of impurities in the base metal, the volume fraction and density of inclusions, and the degree of restraint. The latter effect is in turn dependent on the strength and thickness of the plates being welded. Another factor of importance concerns the dwell time at high temperatures; high energy welding processes thus increase the suscepti-bility to this problem.

Lamellar tearing

An example of lamellar tearing is shown in *Figure 4.13*, in which it is seen that the crack appears to be closely associated with the edge of the HAZ. The horizontal and vertical cracking of the base metal (*Figure 4.13*) is a very typical feature of lamellar tearing. The problem occurs particularly when making tee

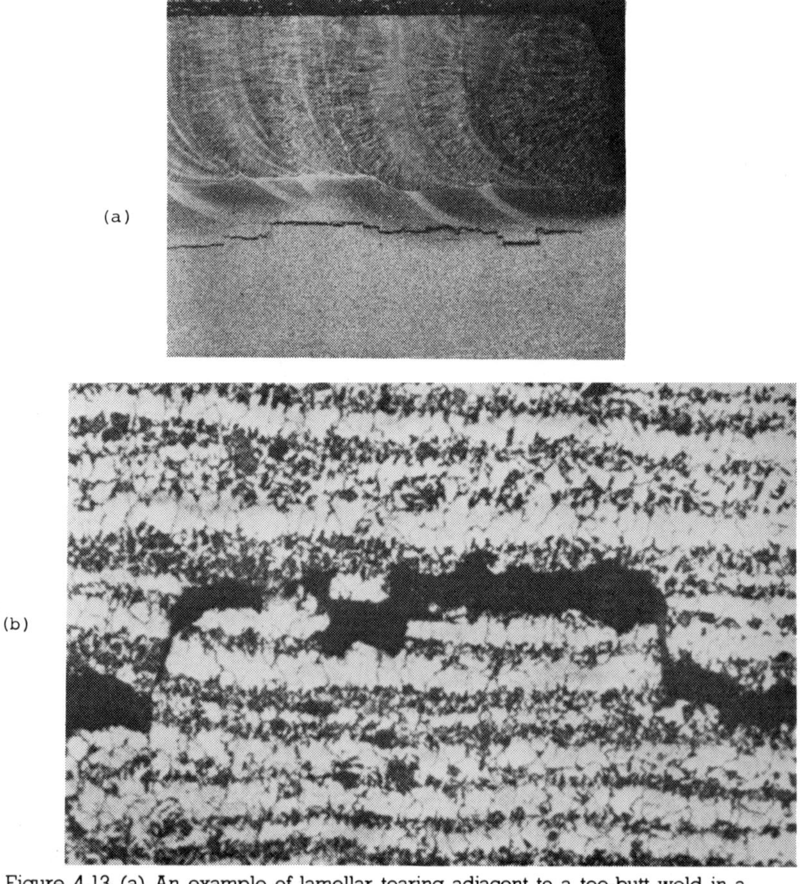

Figure 4.13 (a) An example of lamellar tearing adjacent to a tee-butt weld in a structural steel. After Farrar, J.C.M. and Dolby, R.E., *Lamellar Tearing in Welded Steel Fabrication,* The Welding Institute, 1972. (b) The stepped nature of lamellar tearing in relation to the banded microstructure of the plate. After Colangelo, V.J. and Heiser, F.A., *Analysis of Metallurgical Failures,* John Wiley and Sons, 1974

and corner joints in thick plates, such that the fusion boundary of the weld runs parallel to the plate surface, *Figure 4.14*.

At its simplest, lamellar tearing can occur as a result of a very low through-thickness or short transverse (z) ductility owing to the presence of segregation of elongated (MnFe)(S) inclusions. These arise in the steel processing stage, beginning with segregation bands during solidification of the ingots,

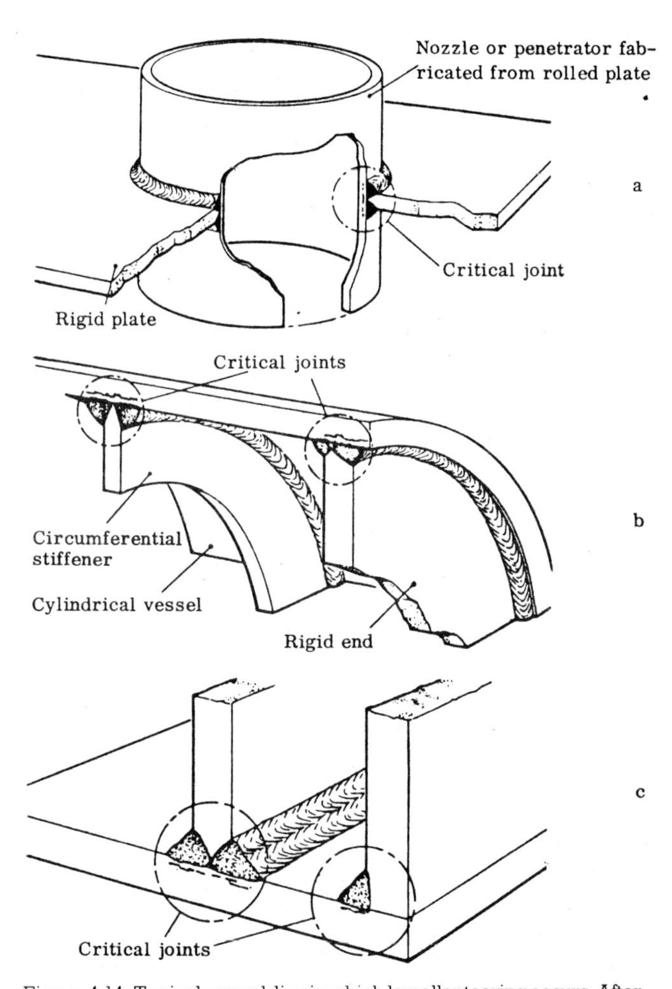

Figure 4.14 Typical assemblies in which lamellar tearing occurs. After *Guide to the Welding and Weldability of C–Mn Steels and C–Mn Microalloyed Steels*, International Institute of Welding, Document IIS/IIW-382-1971, 1971

followed by the rolling and spreading of non-metallic inclusions. The latter can be in the form of large (several hundred microns) FeMnS particles or long stringers of oxides or silicates. Under the action of weld residual stresses, particularly in high restraint geometries, the inclusion–matrix interface ruptures in a number of places; the last stage of fracture causes vertical tearing between planes. This gives the crack its characteristic step-like appearance.

The main factors likely to cause lamellar tearing are thus:

1. Low, short transverse ductility of the parent plate material.
2. A high surface area of planar shaped defects, in the form of flattened out inclusions or stringers.
3. A weld joint configuration which gives rise to high residual tensile stresses in the short transverse direction.
4. The use of thick plates.

On the basis of these factors, lamellar tearing can be avoided by using steels with good transverse properties, such as the Ce-treated grades, which affect the sulphide–iron properties such that the inclusions remain spheroidal even after hot rolling. However, such treated steels are comparatively expensive so that the construction plant often prefers to solve the problem by other means, e.g. by designing to avoid high restraint joints (*see*, e.g. ref. 10).

Another particularly uncomfortable feature of lamellar tearing is that, in some cases, the phenomenon may occur after a certain time delay, taking weeks and even years of service before the cracks appear or failure occurs. It is this aspect that has provoked some researchers to speculate that hydrogen embrittlement may also be involved in this problem. While lamellar tearing appears to be a phenomenon both well documented and avoidable, the exact mechanism by which crack growth occurs is not fully understood as yet. However, certain conclusions may be made, as are discussed below.

Mechanism of lamellar tearing

There are two stages to failure by lamellar tearing:

1. Crack initiation.
2. Crack propagation.

Crack initiation can be explained; it is, however, crack propagation which is less understood.

Crack initiation Crack initiation occurs at the flattened MnS or oxide–silicate stringers by rupture of the inclusion–matrix interfaces[8]. Equation (4.9) governs crack nucleation in this case, and the work of adhesion for MnS–Fe and Al_2O_3–Fe are both close to zero. In order to achieve the fracture stress given by equation (4.9) some plastic deformation of the matrix is needed prior to a cavity opening up at the interface. As a first approximation, the critical strain needed in the matrix, ε_c, to induce cavitation at a hard (non-plastically deforming) particle, is given by equation (4.11)[11], where μ is the shear modulus

$$\varepsilon_c \approx \frac{d}{\mu \bar{b} r} \times \gamma_a \tag{4.11}$$

of the matrix, d is the distance that separates the particles and r is the particle radius. Although this equation was developed for a uniform distribution of spherical particles, as a rough approximation it can usefully be applied to the present case (*see*, e.g., ref. 8). Thus, since in the case of MnS, γ_a is low and r is large (at least for deformation in the short transverse direction), then ε_c is expected to be very low. Similarly, if the density of inclusions in the transverse direction is high, ε_c is further reduced.

Experiments have, indeed, verified low values of ε_c in MnS-steels, an example of which is shown in *Figure 4.15* (*a*). This sequence of micrographs was taken in a scanning electron microscope fitted with a specially instrumented deformation stage; the insert is from the actual load–strain curve of the

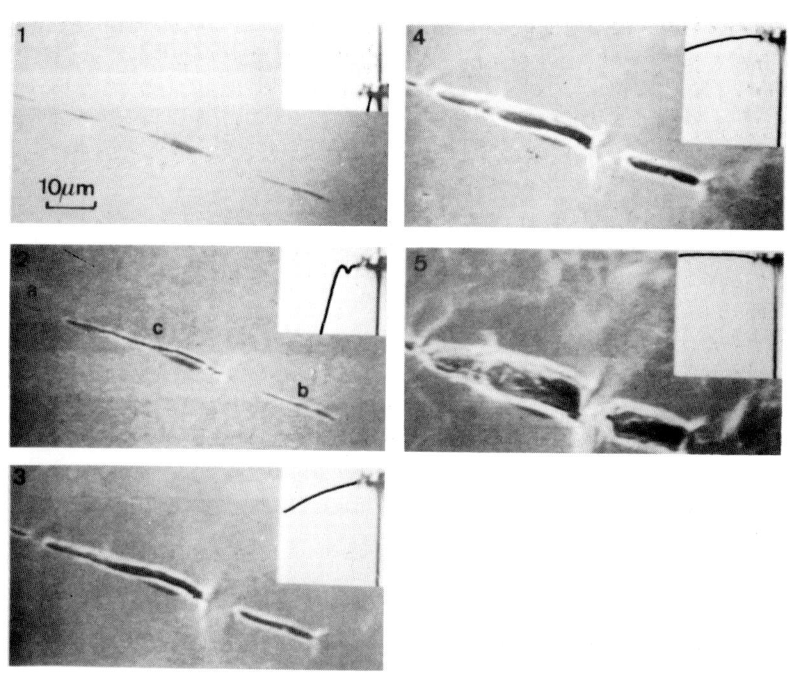

a) cavitation at MnS inclusion at specimen surface

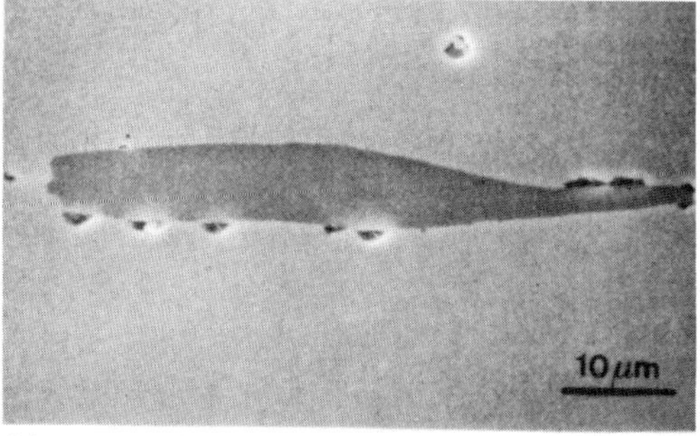

b) cavitation at MnS inclusion below the surface

Figure 4.15 Examples of cavitation at MnS inclusions in a structural steel. (a) A sequence taken *in situ* in the scanning electron microscope using an instrumented deformation stage for recording the load–strain curve (inset). (b) An inclusion below the surface as obtained by sectioning. After Roberts, Lehtinen and Easterling[8]

specimen tested. Cavitation at the MnS inclusion occurs almost exactly at the yield point of the steel. In fact, careful measurements[8] established that at the surface of these steels, $\varepsilon_c \approx 0.008$, while internally (as checked by sectioning) $\varepsilon_c \approx 0.05$, *see Figure 4.15* (*b*). This difference between surface and internal crack nucleation strain is important. In fact, not only do cracks form first at the surface, but the growth of voids also occurs faster at the surface.

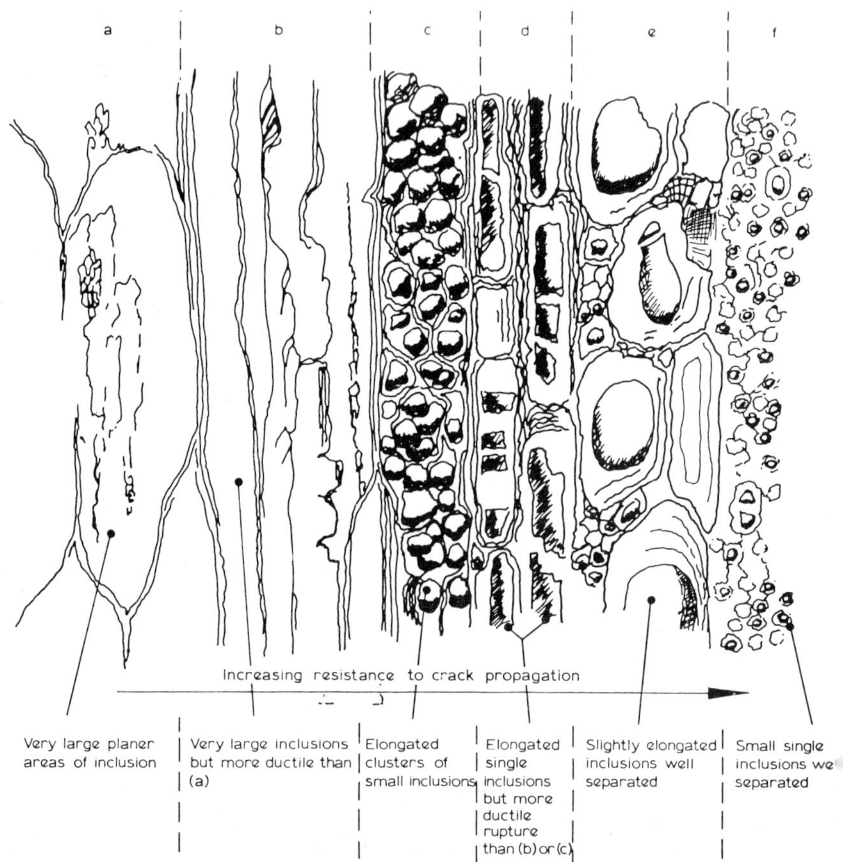

Figure 4.16 Diagramatic illustration of the fracture characteristics associated with lamellar tearing and the relative influence of various types of inclusions. After Farrar, Charles and Dolby[12]

Assuming a ductile matrix, the crack, once formed, does not grow in length if plastic relaxation is allowed to occur at its tip (*see* pp. 156–160). On this basis the crack *widens* plastically, the cavity volume, V_c, increasing with strain as in the relationship (4.12)[11], where V_p is the particle volume. This

$$V_c = V_p (\varepsilon - \varepsilon_c) \tag{4.12}$$

simple relationship has been verified experimentally[8]. The influence of a free surface in accelerating the nucleation and growth of cavities around inclusions depends on the magnitudes of the compressive stresses that develop perpendicular to the tensile axes during deformation; these must be especially large for inclusions of high aspect ratio when loaded perpendicularly to their long dimension.

Metallurgical examination of lamellar tearing initiation confirms many of the characteristics expected on the basis of the above discussion (*see*, e.g., ref. 12). The tendency, for example, for cracks to initiate near free surfaces has been verified. Fractographic studies have confirmed the potency of the large, elongated, coplanar MnS inclusions, but show that rows of small oxide–silicate inclusions can also be effective in initiating and spreading a crack. A diagram that summarizes some of these fractographic observations and illustrates the relative potency of the various types of inclusion in initiating cracks is shown in *Figure 4.16*. The ductility of the matrix, in contrast to the complete lack of

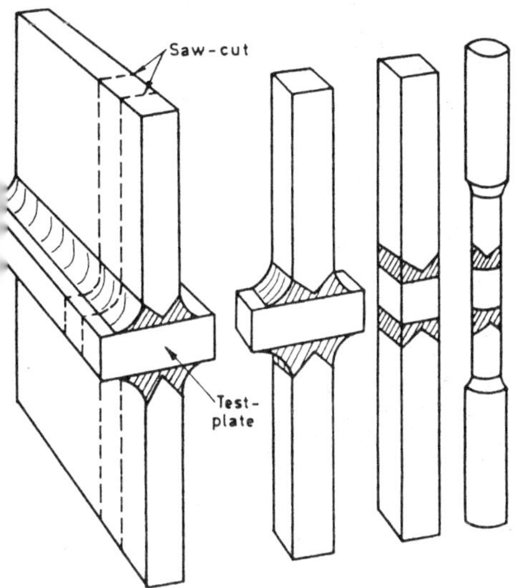

Figure 4.17 Welded test-bar for z-direction testing of steel plates showing the stages of specimen preparation. After Wold, G. and Kristoffersen, T., Development of a method for measuring susceptibility of steel plate to lamellar tearing: preliminary results, *Sveiseteknikk*, **27**, 33, 1972

ductility of the inclusions, is a notable feature of these observations. The dependence of the volume fraction of local inclusions on strain-to-fracture has been well catalogued, and this is, in fact, the basis of one of the standard tests for measuring susceptibility to lamellar tearing. This experiment is illustrated in *Figure 4.17*, (*see also* ref. 13).

Crack propagation Crack propagation to link cracks from inclusions lying in different planes appears to occur by vertical and angled steps (*see Figure 4.13*). These steps are thought to occur by a ductile tearing of the matrix between expanded cavities, *see*, e.g., *Figure 4.15*.

While the onset of lamellar tearing is considered to be mainly dependent upon the local volume fraction and surface area of inclusions in the plane of the plate, the close association of the phenomenon with the *edge* of the HAZ in *thick* plates is also significant. Indeed, the matrix plasticity at the crack tip is thought to be affected by *strain ageing*, a phenomenon known to occur at the edge of the HAZ, particularly in high energy or multi-run welds (*see* pp. 141–

143 and *Figure 4.5*). Strain ageing hardens the matrix and thereby decreases its ductility. This, in conjunction with the high, elastic residual stresses associated with thick plates, acts to reduce the critical size of the crack for rapid crack propagation.

That lamellar tearing is sometimes a time-dependent process has at least two possible explanations. One possibility is that hydrogen embrittlement contributes to reduction of the ductility of the matrix. The other is that additional stresses in service add to the residual stresses of the weld. Deco-

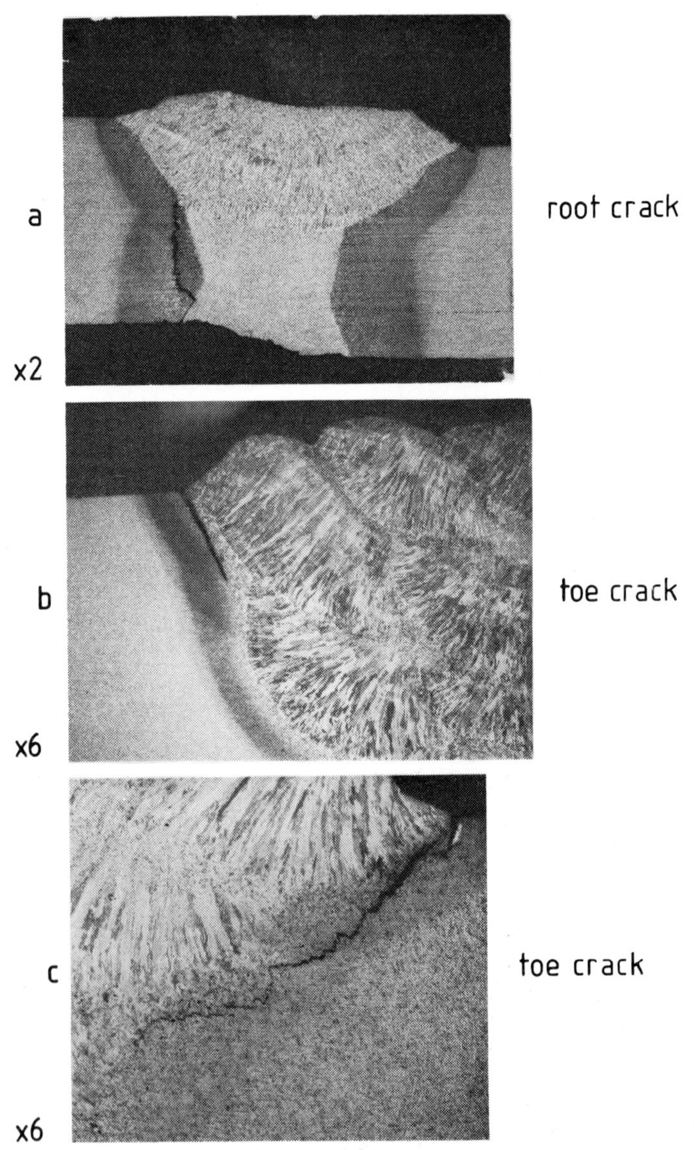

Figure 4.18 (a)–(c) Examples of hydrogen or cold cracking in the heat-affected zone of welds. All micrographs are by courtesy of ESAB, Gothenburg (reduced by two-fifths in printing)

hesion of inclusions due to weld residual stresses has the effect of reducing the stress level. Indeed, this and the fairly good matrix plasticity may be sufficient to arrest crack propagation. Hence, additional tensile stresses that arise in service may tend to be more severe with increased decohesion, because of the effective reduction in cross-sectional area of the material and increased stress concentration at the voids. Finally, if the service stresses are of a cyclic nature there may be a tendency to harden the matrix by increasing dislocation density and this effect would be largest around inclusions.

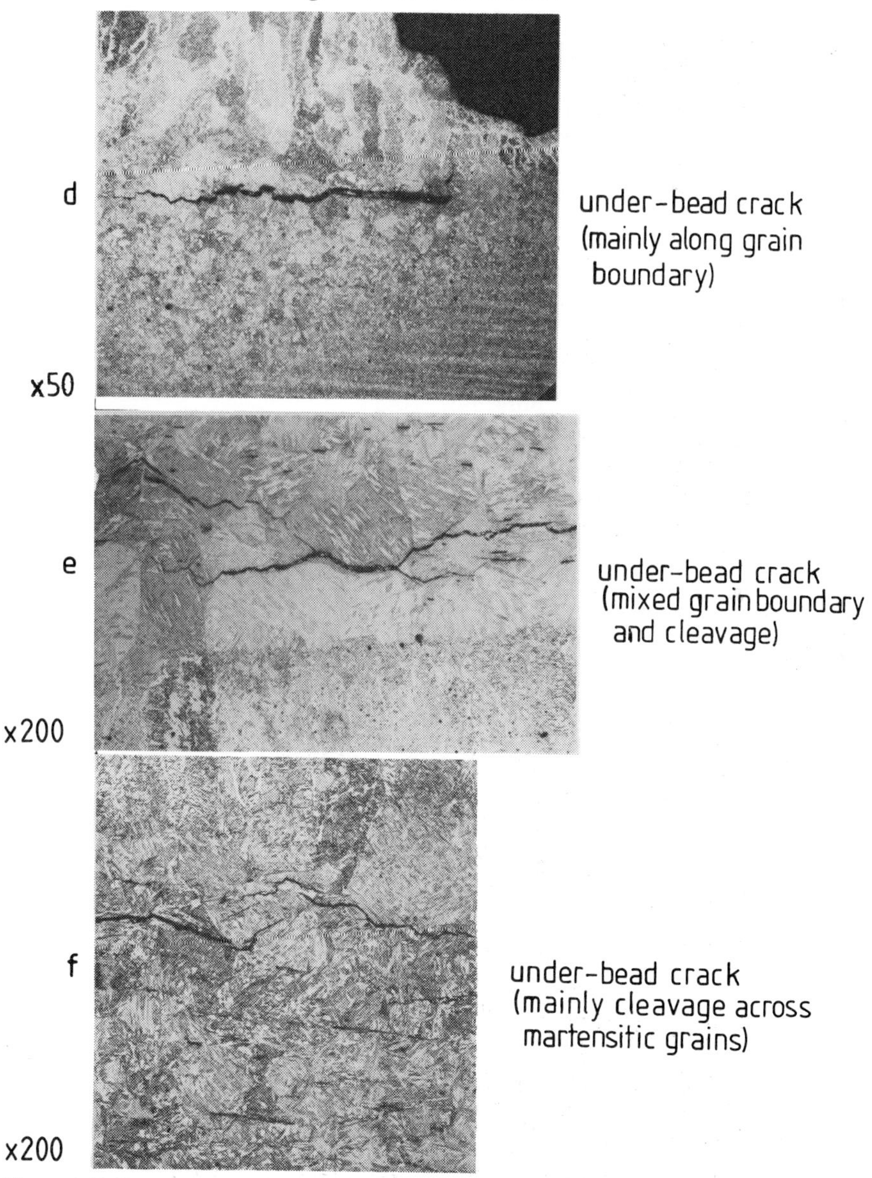

Figure 4.18 (d)–(f) Examples of hydrogen or cold cracking in the heat-affected zone of welds. All micrographs are by courtesy of ESAB, Gothenburg (reduced by two-fifths in printing)

Cold cracking

Cold cracking, or hydrogen-induced cracking, is perhaps the most serious and least understood of all weld cracking problems. Examples of cold cracking are illustrated in *Figure 4.18* which shows how these cracks can typically appear in the HAZ (underbead; root crack), as a toe crack (from the surface), or in the weld metal.

The various factors that influence hydrogen cracking are well documented (*see,* e.g. ref. 14). Hydrogen is introduced into the weld during the welding process (*see* pp. 9–17, *Figures 1.4* and *1.9*), from the atmosphere, from hydrocarbons on the plate being welded, or from using a damp electrode flux.

Prior to crack initiation there is usually a time delay, or *incubation time,* and crack propagation occurs in a slow, jerky mode, with further incubation times between steps in the crack growth. The crack can be both trans- and inter-crystalline in character, but mainly follows, in this case, prior austenite grain boundaries. The initiation of cold cracking is particularly associated with notches, such as the toe of the weld, or microstructural inhomogeneities which exhibit sudden changes in hardness, such as slag inclusions, martensite–ferrite interfaces or even grain boundaries. In common with most other crack growth phenomena, cold cracking is accentuated in the presence of high restraint weld geometries and matrix hardening.

The gradual loss of notch toughness of samples charged with hydrogen has been well characterized experimentally. An example of the reduction in notched tensile strength of various steels as a function of time is shown in *Figure 4.19.* Structures that contain twinned martensite are usually found most susceptible.

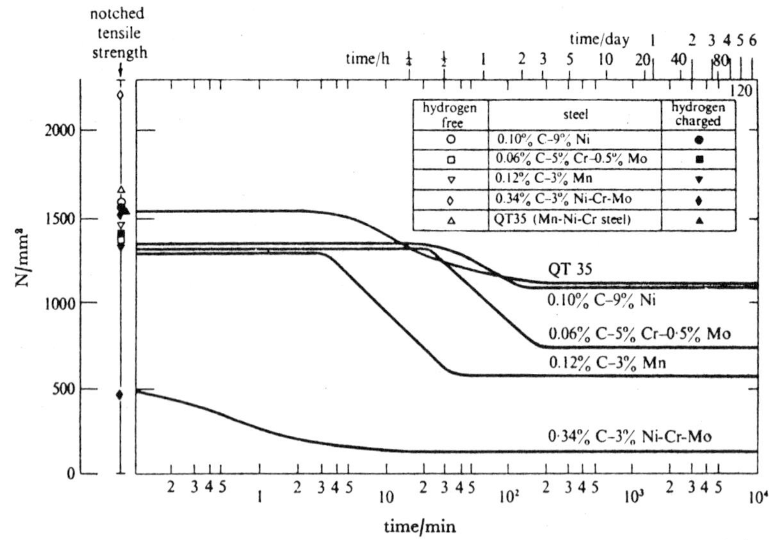

Figure 4.19 Constant load rupture strength of different hydrogen-charged steels in the quenched condition to simulate HAZ microstructures. After Baker, R.G., Weldability and its implications for materials requirements *Rosenhain Centenary Conference,* p. 129, eds Baker, R.G. and Kelly, A., Royal Society, 1975

Cold cracking is basically dependent upon three mutually interactive factors, these being:

1. The presence of hydrogen (even very small amounts, i.e. parts per million).
2. High residual stresses.
3. A 'susceptible' microstructure.

Role of hydrogen

Compared with other interstitials, hydrogen diffuses much more easily through b.c.c. iron. This is demonstrated in *Table 4.2* which compares the diffusivities of hydrogen with carbon and nitrogen at different temperatures. At 20 °C hydrogen atoms are estimated to be capable of jumping at speeds *ca.* 10^{12} times faster than C and N! Indeed, this has lead Fast[15] to suggest that hydrogen behaves more as a *proton* than as an atom. On the other hand, the maximum solubility of hydrogen in iron at the melting temperature is only *ca.* 30 ppm. This suggests that the diffusion of hydrogen atoms in iron must occur practically *independently of stress*. There is, in fact, experimental evidence that this is so[16], and the implications of this are returned to later.

TABLE 4.2 Diffusivities of hydrogen, carbon and nitrogen in b.c.c. iron (after Fast[15])

Temperature (°C)	D (*H*) $mm^2\,s^{-1}$	D (*C*) $mm^2\,s^{-1}$	D (*N*) $mm^2\,s^{-1}$
20	1.5×10^{-3}	2.0×10^{-15}	8.8×10^{-15}
100	4.4×10^{-3}	3.3×10^{-12}	8.3×10^{-12}
200	1.0×10^{-2}	1.0×10^{-9}	1.7×10^{-9}
300	1.7×10^{-2}	4.3×10^{-8}	5.3×10^{-8}
400	2.5×10^{-2}	5.9×10^{-7}	6.0×10^{-7}
500	3.3×10^{-2}	4.1×10^{-6}	3.6×10^{-6}
700	4.9×10^{-2}	6.1×10^{-5}	4.4×10^{-5}
900	6.3×10^{-2}	3.6×10^{-4}	2.3×10^{-4}

An interesting feature of hydrogen diffusivity measurements in ferritic steels is the wide scatter of results at temperatures below *ca.* 200 °C, *Figure 4.20*. Indeed, the diffusivity measurements given in *Table 4.2* for hydrogen must be considered *ca.* 100 times too high on the basis of *Figure 4.20*. For this reason it is usual to assume two diffusivity coefficients for hydrogen diffusion in α–Fe based on experimental measurements[15], one at 200 °C and above given by equation (4.13) and the other for below 200 °C given by equation (4.14).

$$D = 0.14 \exp\left(-\frac{13400}{RT}\right) \quad \text{mm}^2\,\text{s}^{-1} \tag{4.13}$$

$$D = 12 \exp\left(-\frac{32700}{RT}\right) \quad \text{mm}^2\,\text{s}^{-1} \tag{4.14}$$

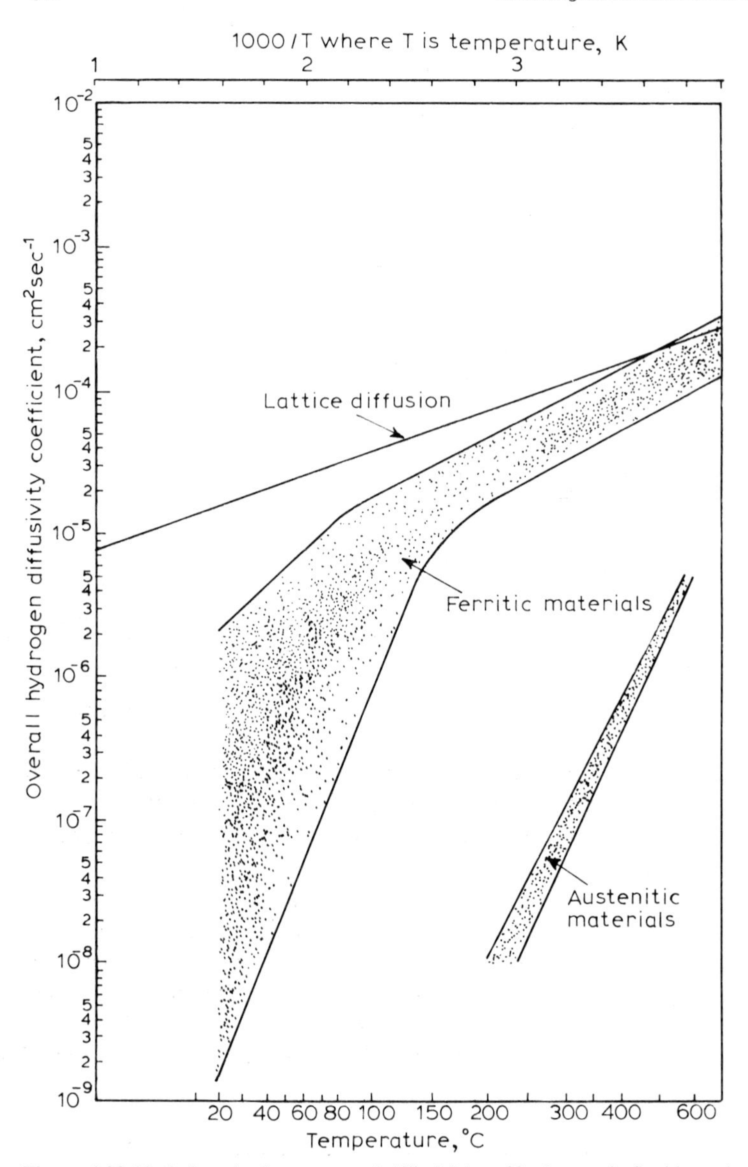

Figure 4.20 Variations in the measured diffusivities of hydrogen in ferritic and austenitic steels. After Coe, F.R., *Welding Steels without Hydrogen Cracking*, The Welding Institute, 1973

The reasons for the diffusivity reduction and increased scatter below 200 °C are not completely understood, but there is good evidence that hydrogen becomes trapped at various defects where it transforms to molecular hydrogen. Precipitation is thought to occur at most defects available, including inclusion–matrix interfaces, internal cavities or cracks, grain boundaries and dislocations. Calculations of the volume fraction of hydrogen sinks, or voids, needed to account for these changes in diffusivity have been carried out, assuming the presence of 10 ppm of hydrogen[17], *Figure 4.21*, which tend to substantiate the

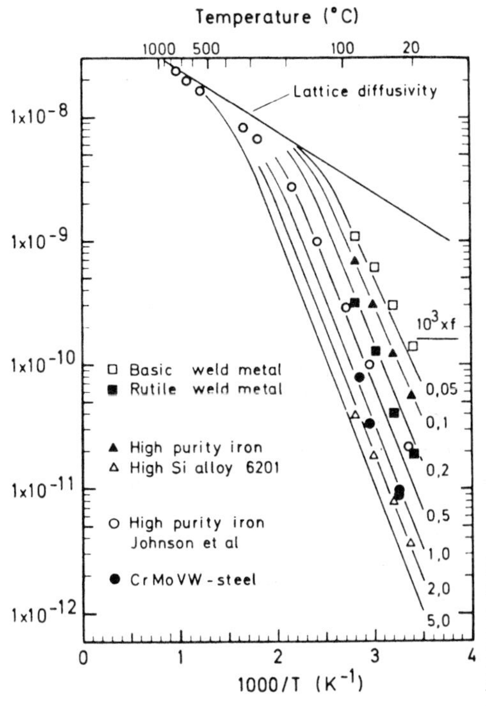

Figure 4.21 Calculated values of apparent diffusivity D', as a function of temperature and relative void volume, f. After Andersson, B.A.B., Diffusion and trapping of hydrogen in a bead-on-plate weld, *Journal of Engineering Materials and Technology*, **102,** 64, 1980

'sink' model. Since the hydrogen precipitated as H_2 is no longer in solution, the lower diffusivities measured below *ca.* 200 °C have been attributed to some sort of controlled re-solution of hydrogen when diffusing from one sink to another. What controls this redistribution, however, is not clear.

The main source of hydrogen in welding arises from the dissociation of water vapour in the welding arc, but its exact origin depends on the type of process employed. For example, as little as 1 cc per 100 g can be of significance when welding hardened steels. The distribution of hydrogen as a function of the movement of the arc is shown schematically in *Figure 4.22*.

As shown in *Figures 1.7* (p.12) and *4.20*, a decrease in hydrogen solubility

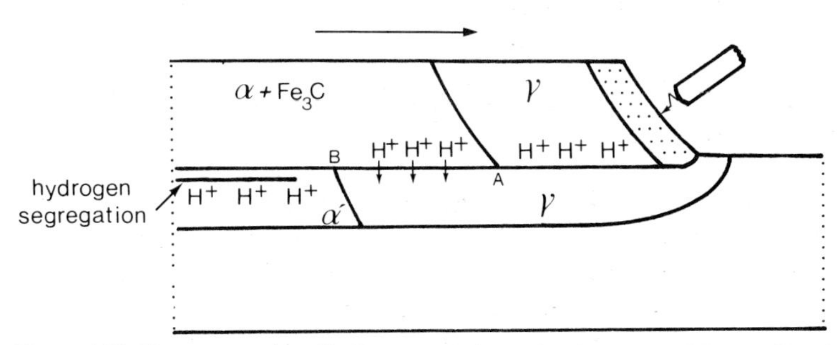

Figure 4.22 The propagation of hydrogen relative to the movement of the arc. A and B refer to transformation fronts in weld metal and base material. After Granjon, H., Cold cracking in welding of steels, *International Symposium on Cracking and Fracture in Welds (Conference Proceedings)*, Japan Welding Society, Tokyo, **IB,** 1.1, 1971

occurs during the $\gamma \rightarrow \alpha$ transformation which corresponds to an increase in diffusivity. Because of the high mobility of hydrogen it should be easy for hydrogen diffusion to accommodate differences in solubility associated with the austenite–ferrite transformation. Thus, as the transformation proceeds the austenite becomes progressively more enriched in hydrogen. This may be why cold cracking is so sensitive to the presence of martensite, since the austenite–martensite transformation occurs at the lowest temperatures and originates from the most hydrogen-enriched austenite. It has also been found that the presence of twinned (high carbon) martensite is more dangerous than lath (low carbon) martensite. This could well be explained by the fact that twinned martensite has a lower M_s temperature, corresponding to a larger dilatation due to the transformation. The higher carbon content of this

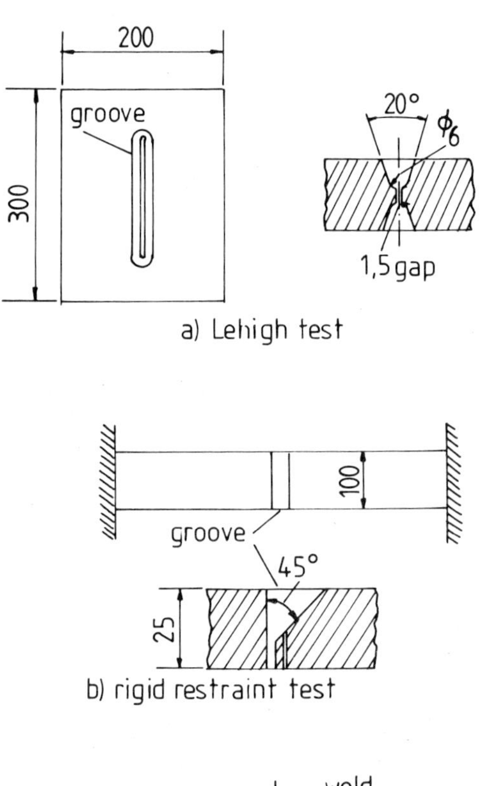

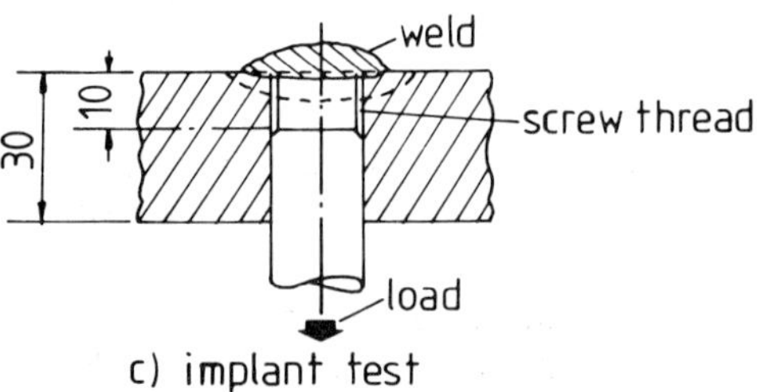

Figure 4.23 Examples of hydrogen-cracking tests for welds. The Tekken test is similar to the Lehigh test except that it is grooved on one side only. All dimensions are in mm.

martensite also increases its hardness. On this basis, the martensitic transformation is likely to be associated with a relatively high hydrogen enrichment at defects such as lath-packet boundaries or twinned plate boundaries.

Role of stress

The presence of residual stresses in welds is discussed in some detail on pp. 33–45, where it is shown that in high restraint welds tensile stresses of the order of the yield strength of the metal are possible.

The role of stress has been studied in various hydrogen cracking tests, e.g. the *Lehigh, Tekken, rigid restraint* and *implant* tests, *see Figure 4.23*. The Lehigh and Tekken tests probably more closely resemble the real weld situation, particularly in butt welding, since the restraint factor and stress

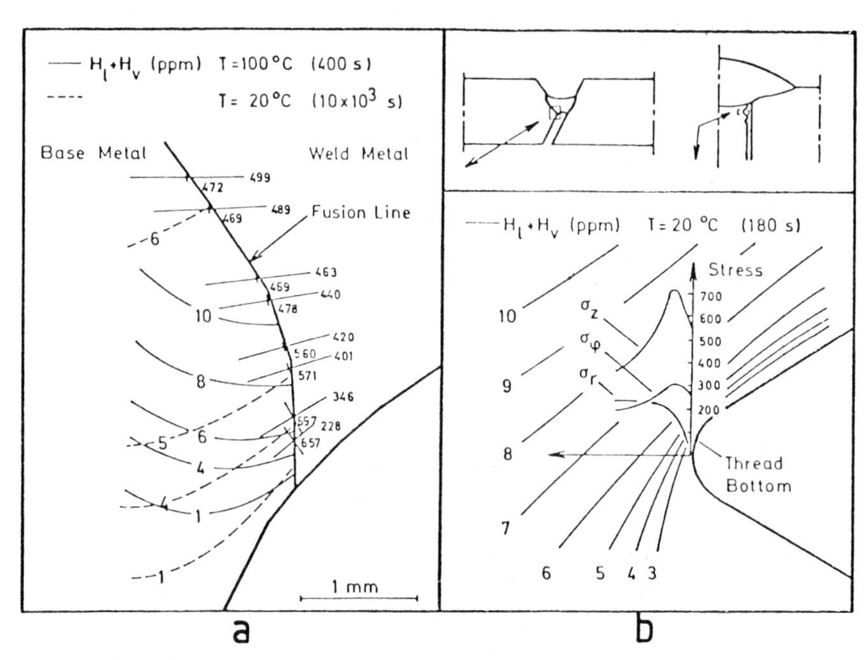

Figure 4.24 Calculated stresses (in MPa) and residual hydrogen (ppm) distributions in (a) Tekken and (b) implant test specimens. The contours refer to hydrogen concentrations in ppm (low numbers) and stresses in MPa (high numbers). After Andersson[18]

distributions are more like those imposed in welding. However, the implant test, based on a notch in the form of a screw thread positioned at the HAZ, is also considered to be a reliable method of ranking the susceptibility of various steels to hydrogen cracking (*see,* e.g., ref. 17).

A study of the stress and hydrogen distributions at the Tekken test root and implant thread root of quenched and tempered steels has been carried out by Andersson[18] using an FEM analysis and his results are shown in *Figure 4.24*. The hydrogen distributions shown correspond to post-welding times of 400 s and 10^4 s at temperatures of 100 °C and 20 °C, respectively, and are calculated assuming a stress-dependent diffusion coefficient of hydrogen from an initially uniform hydrogen field. The maximum acting stresses in the two tests are rather similar, although in the case of the Tekken test the maximum stress component is the longitudinal stress, whereas in the implant test the maximum stress is that acting perpendicular to the fusion line or specimen axis. In neither case, however, is there any tendency for the hydrogen to be strongly attracted to the regions of maximum tensile stress.

In another FEM analysis, Andersson[18] examined the stress and hydrogen distributions ahead of a relatively sharp notch which corresponded to a blunt crack, and this result is shown in *Figure 4.25*. The crack opening displacement,

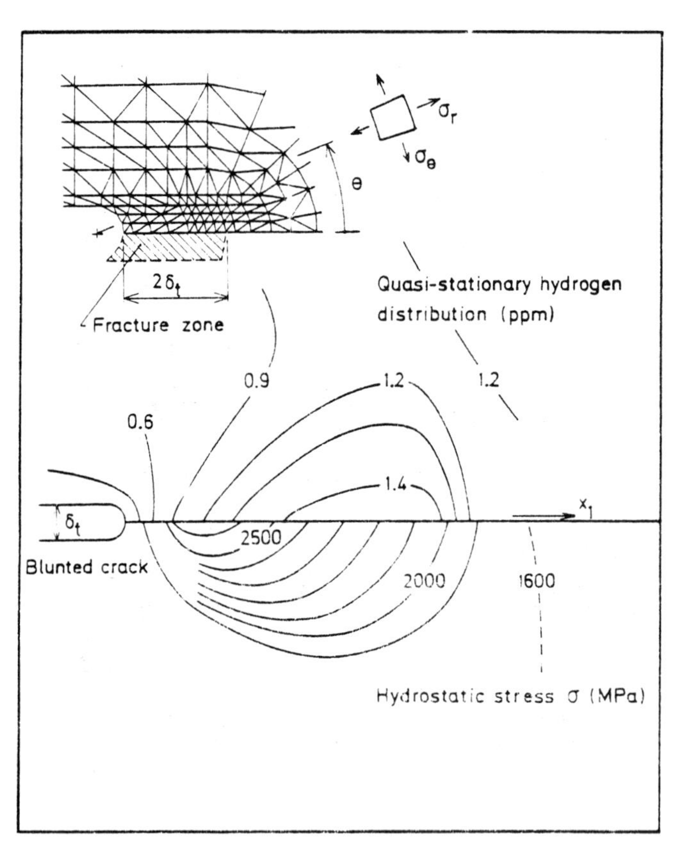

Figure 4.25 Quasi-stationary hydrogen distribution and stationary hydrostatic stress around a blunted crack moving with a constant velocity of 5×10^{-8} m s^{-1}. The size of the FEM mesh is 30×60 μm. The crack opening displacement is $\delta = 3.1$ μm. After Andersson[18]

δ_{IC}, in this case was 3.1 μm, and the analysis assumed a crack velocity of 5×10^{-8} m s^{-1}. This crack growth rate was deduced from acoustic emission measurements of hydrogen cracking in real welds[19]. The area covered in *Figure 4.25* is 30×60 μm which corresponds to about a grain diameter in the grain growth zone of the quenched and tempered steel studied. This interesting result shows that a very slight maximum in hydrogen distribution occurs ahead of the moving crack, although it does not correspond to the triaxial stress maximum.

The theoretical analysis described above indicates that stress only weakly influences hydrogen distribution ahead of cracks, and thus tends to support experiment[15, 16]. Studies of the microstructure in implant specimens of micro-alloyed steels prior to fracture have revealed a fairly uniform distribution of cracks at distances of several notch radii ahead of the thread roots, i.e. in areas that do not correspond to stress maxima, as shown in *Figure 4.26*. Closer

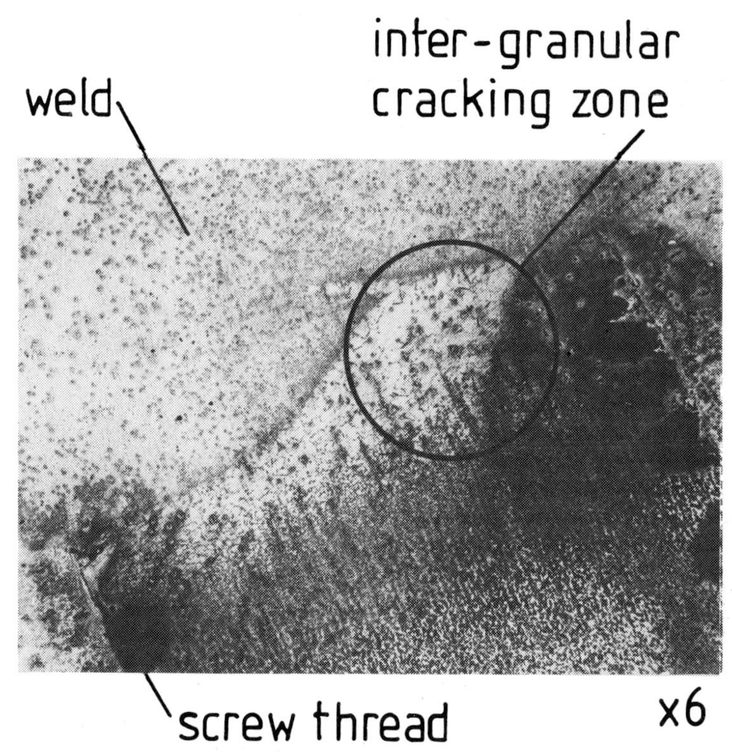

Figure 4.26 Crack initiation in an implant test of a structural steel prior to fracture. The cracks, mainly confined to within the circle marked, appear to be predominantly inter-granular. Courtesy of P–A. Morin and U. Wuopio, University of Luleå

investigation of these cracks revealed that many were intergranular, decorating prior austenitic grain boundaries, and that practically all lay parallel to the fusion line, i.e. perpendicular to the tensile stress. Final fracture in these specimens appeared to spread from within the sample to near the thread root-surface, predominantly in an intergranular mode.

Role of microstructure

It is generally found that cold cracks initiate in the grain growth zone of the HAZ. As discussed in detail earlier, the increase in grain size effectively decreases the transformation temperatures so that in high C_{equiv} steels the volume fraction of the lower temperature transformation products, such as martensite, bainite or Widmanstätten side plates, is increased. The high dislocation densities associated with these products, together with fine carbide particle hardening, are thus likely to result in a hard, low ductility matrix. Another effect of larger prior-austenitic grain size is to increase the amount of segregation at boundaries, and this may be revealed in TEM in the form of grain-boundary precipitates.

The importance of grain boundaries and second-phase particles as sinks for hydrogen appears to be well established. Indeed, this subject has recently received a great deal of attention in the literature, and accelerated decohesion and void growth at non-metallic inclusion interfaces have been reported for, e.g., stainless steels[20], pearlitic steels[21] and spheroidized, plain carbon steels[22, 23], to quote just a few. The effect of hydrogen on the ductile fracture of

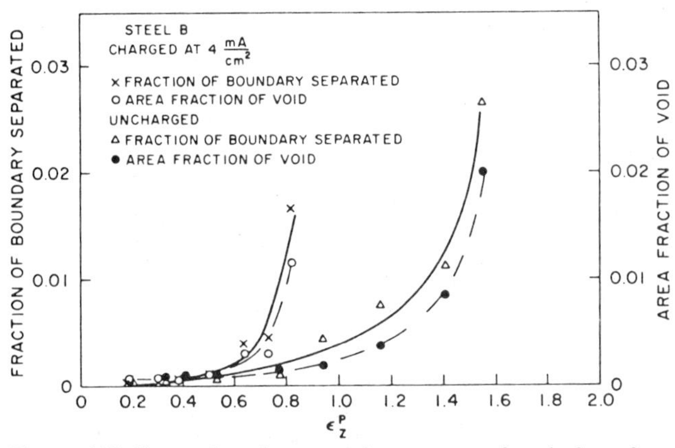

Figure 4.27 Comparison between the amounts of grain boundary rupture in hydrogen-charged (left-hand curve) and uncharged (right-hand curve) samples. After Cialone and Asaro[23]

steel is evidently to reduce ductility as measured by a decrease in reduction of area at fracture, in spite of the fact that the flow characteristics of the metal (i.e. dislocation mobility) appear to be unchanged, at least for the relatively mild hydrogen doses associated with welding. Under more severe charging conditions, it is thought that hydrogen may even affect flow conditions by exerting high pressures at internal voids.

It is interesting that decohesion of cementite particles can occur in the presence of hydrogen, since normally the cementite–ferrite interface is considered to be very well bonded, i.e. its work of adhesion is high (for good evidence of this, *see,* e.g., ref. 24). How, then, does hydrogen reduce the work of adhesion so effectively? One explanation is evidently[15] that contact between hydrogen atoms and cementite brings about reaction (4.15). Should this

$$Fe_3C + 4[H] \rightarrow CH_4 + 3Fe \quad (4.15)$$

reaction occur, only small amounts of hydrogen would be needed to produce thin layers of CH_4 and to reduce γ_a to zero. Since carbide formation often

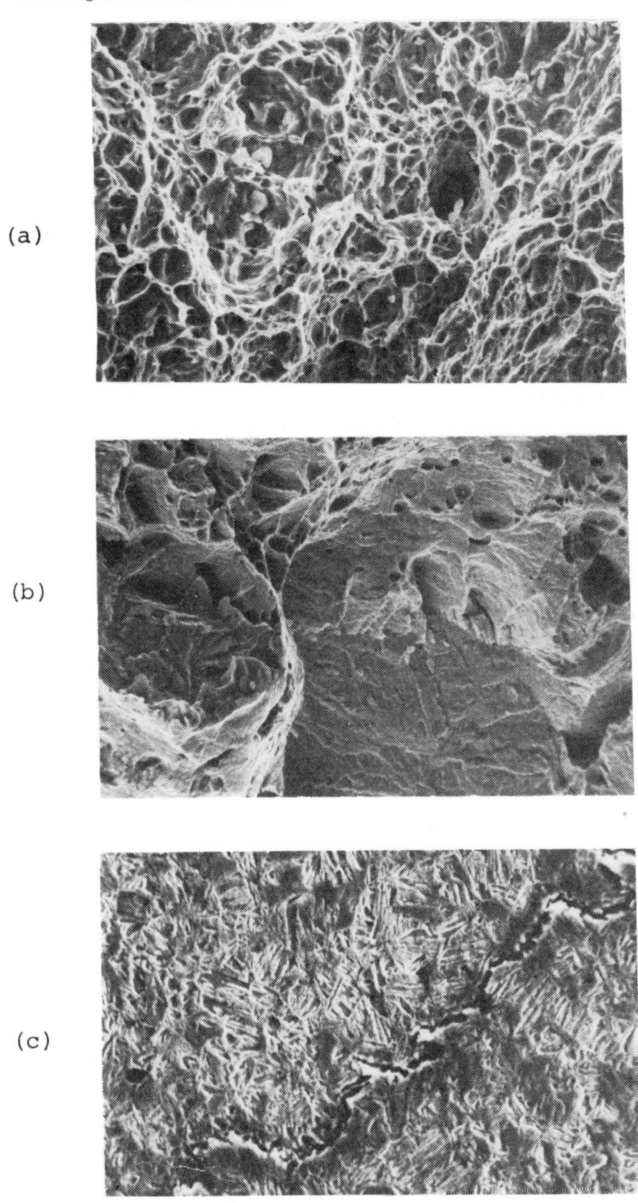

Figure 4.28 Comparison between the ductile fracture characteristics of a bainitic steel (a), and the predominantly cleavage fracture appearance of the same steel charged with hydrogen ((b) and (c)). Courtesy of A. Lindfors, University of Luleå (× 1000; then reduced by one-fifth in printing)

occurs preferentially at grain boundaries, and grain boundaries are likely sinks for hydrogen during cooling, this reaction appears to provide a feasible explanation for the grain boundary rupture of cold cracking. Reactions with the metallic components of inclusions may also occur, e.g. with Ti or Nb, to produce brittle metal hydride films at the interfaces.

Void growth in grain boundaries in steels that contain grain boundary carbides has been studied, *see* ref. 23 and *Figure 4.27*. The fraction of grain boundary rupture corresponds closely to the total void area in both charged and uncharged samples, and the ductility of the charged samples is considerably less than that of the uncharged specimens.

In specimens that contain higher hydrogen charging a phenomenon known as *'fisheyes'* is sometimes observed at fracture surfaces. These are interpreted as circular areas of quasi-cleavage fracture centred on a large inclusion (e.g. MnS) or clusters of small inclusions (e.g. silicates or Al_2O_3 particles). Indeed, even in samples that contain relatively low levels of hydrogen it is found that the tendency for brittle cleavage fracture is increased over that of uncharged samples.

Examples of this transition in a bainitic steel are shown in *Figures 4.28* (*a*)and (*b*) which compare the fracture surfaces before and after charging with hydrogen. The fracture path appearance in (*b*) is a mixture of cleavage and ductile fracture and shows that the fracture path is constantly forced to change direction. This is illustrated in (*c*) which shows the zig-zag nature of the cleavage fracture-path, corresponding to easy propagation along, and less easy propagation across, the bainite laths.

In summary, a microstructre which is considered *susceptible* to cold cracking contains:

1. Fairly hard grains consisting of, e.g., martensite or bainite.
2. A coarse prior-austenite grain size.
3. Slag inclusions, or coarse carbide particles that decorate grain boundaries.
4. A high dislocation density.

Mechanism of cold cracking

To explain the phenomenon of cold cracking, clarification of a number of well established characteristics is required. Assuming certain minimum levels of tensile stress, hydrogen contamination and susceptible microstructure, these characteristics include:

1. Crack growth is a time dependent process.
2. Crack growth occurs extremely non-uniformly, propagating rapidly between lengthy periods of arrest.
3. Crack propagation is often of a semi-brittle, cleavage nature, although in coarse grained material in which the grain boundaries are decorated with particles, it can be (and often is) intergranular.
4. Hydrogen agglomeration is only weakly dependent upon stress.

Most analytical models for estimating the crack growth velocity in cold cracking assume that diffusion of hydrogen to the region of triaxial stress ahead of the tip is the rate-controlling process. Recent measurements show only a weak dependence of hydrogen diffusion on stress, however, so this may not be a very realistic model. Nevertheless, as shown in *Figure 4.25*, a slight increase in hydrogen level may occur ahead of an arrested crack and, even if this does not coincide with the highest triaxial stress, it could be sufficient to nucleate microcavities, e.g. at dislocation pile-ups ahead of the arrested crack[18]. In this case the criterion for crack growth is that the microcavities can link up and join the main crack. The maximum stress ahead of the microcavity is (from equation (4.4)) given by equation (4.16), where c_1 is the crack length of the

$$\sigma = \sigma_a (\pi c_1)^{1/2} \tag{4.16}$$

microcavity. This cavity is conjectured to grow as more hydrogen condenses between the two crack surfaces. Crack propagation ultimately occurs under conditions given by equation (4.17)[25], where C_H is the average hydrogen concentration per unit area and F_m is the cohesive force of the lattice. Note that during the time that hydrogen condenses ahead of the crack, the value of the right hand side of equation (4.17) is decreasing. At some critical hydrogen

$$\sigma_a(\pi(c + c_1))^{1/2} \leq F_m(C_H) \tag{4.17}$$

level, the cohesive lattice strength is envisaged to fall below the fracture strength and a microcrack forms, as shown schematically in *Figure 4.29*. This microcrack then joins up with the main crack and the whole procedure is repeated.

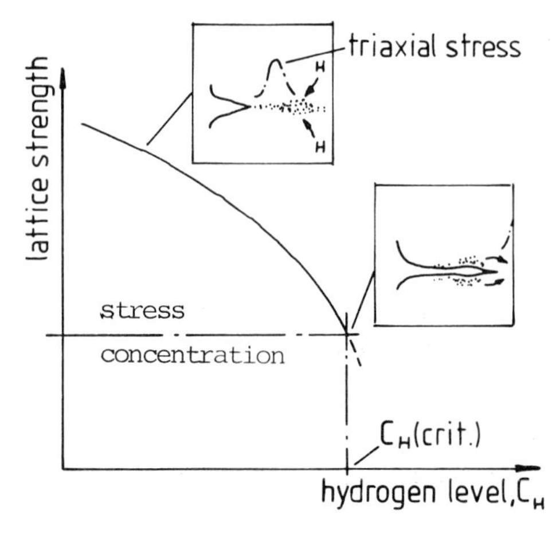

Figure 4.29 Schematic diagram to illustrate equation (4.17), which shows that as the hydrogen level ahead of a crack reaches a certain critical level, the cohesive lattice strength falls below the stress concentration at the notch, thus allowing a microcrack to nucleate. This microcrack is then envisaged to join up with the main crack as shown

An alternative mechanism for crack growth has been proposed[26, 27], in which the arrested crack acts as a sink for hydrogen atoms which become absorbed at the crack tip. Thus, the surface energy of the crack is reduced such that, at some critical value of surface energy, crack propagation may occur. In terms of the Griffith criterion, the fracture stress is then given by equation (4.18) where γ_s' is the reduced surface energy of the matrix.

$$\sigma_f = \left(\frac{2E\gamma_s'}{\pi c}\right)^{1/2} \tag{4.18}$$

This essentially brittle fracture criterion can be adopted to explain crack growth in grain boundaries by a mechanism illustrated in *Figure 4.30.* It is conjectured that the crack is able to move along the hydrogen-contaminated boundary by the expansion of cracks nucleated at particles situated on the boundaries. The fracture criterion is governed by an amended fracture stress, σ_f', equation (4.19), where γ_{gb}' is the hydrogen-contaminated grain boundary

$$\sigma_f' = \left[\frac{2E\,(\gamma_s - \gamma_{gb}')}{\pi c}\right]^{1/2} \tag{4.19}$$

energy, and is probably larger than the clean grain boundary energy, γ_{gb}. Thus, the effect of contamination of the grain boundary is to reduce σ_f'. The advantage of this model is that little stress-dependent hydrogen diffusion is needed, since the grain boundaries have already become contaminated with hydrogen during cooling of the weld. Furthermore, this contamination is highest in the coarse-grained austenite because of the lower sink area. This

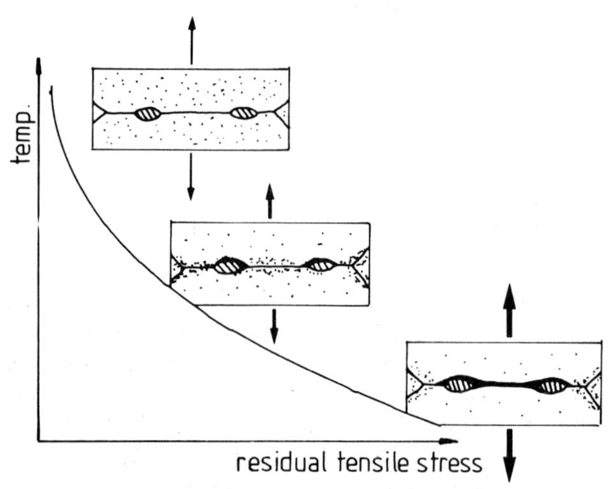

Figure 4.30 The mechanism of hydrogen-induced grain boundary cracking, in which the uniform hydrogen distribution at high temperatures changes at lower temperatures as hydrogen diffuses to the various sinks which initially causes particle decohesion and then grain boundary rupture

model can even by applied to crack growth along, e.g., bainite or martensite lath-packet boundaries (also likely hydrogen sinks), as observed, e.g., in *Figure 4.28* (*c*). The incubation time characteristic is explicable in terms of the time taken for cracks to form at particles under the influence of stress and hydrogen contamination and then join up along the boundaries. The criterion in that case for crack formation at particles is as for equation (4.9), i.e. equation (4.20), where γ_a is the work of adhesion of the particle–matrix interface (*see*

$$\sigma_f^p = \frac{2\gamma_a}{b} \tag{4.20}$$

equation (4.10)). However, as discussed above, γ_a is likely to be close to zero in this case, and hence σ_f^p is very small.

The mechanism of 'fisheye' formation at fracture surfaces is generally thought to be due to high molecular hydrogen gas pressures building up at an inclusion interface. Alternatively, some inclusions, e.g. the basic silicates, may have a higher solubility for hydrogen and this may also give rise to conditions that cause fisheye fracture. *Figure 4.31* illustrates how a fisheye may develop at a MnS inclusion. The magnitudes of pressure build-up likely under weld conditions are difficult to estimate, particularly in commercial steels where the

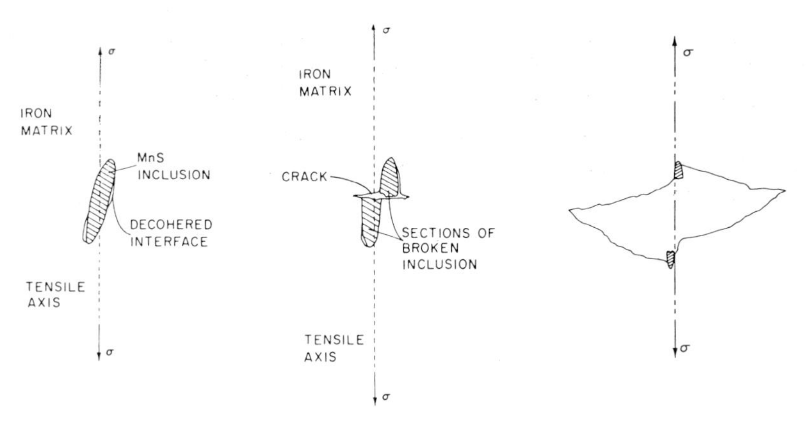

Figure 4.31 Possible mechanism for the formation of 'fisheyes' in conjunction with hydrogen-initiated decohesion at MnS inclusions. After Cialone and Asaro[23]

total void volume is dependent on so many variables. In any case, the internal pressures exerted are thought to be additive to the triaxial tensile stresses which act at the cavity edge and that a Griffith fracture criterion for brittle fracture can therefore be achieved. Reduced surface energy due to hydrogen absorption would further contribute to brittle failure. It seems likely that toe cracks may also be aided by this sort of fracture, inclusions in the vicinity of the toe being convenient sinks for hydrogen escaping to the surface.

In summary, none of the theories discussed can explain all the characteristics of cold cracking and any theory has to be highly speculative at this time. Perhaps the two most critical aspects are the effect of stress on hydrogen diffusion and the movement of hydrogen between defects. Assuming that hydrogen atoms diffuse as protons, i.e. with little or no self (misfitting) stress, they would be practically independent of the influence of external stress. Thus, expansion of the lattice by triaxial tensile stresses appears only weakly to attract hydrogen. On the other hand, the lattice solubility appears to be extremely low since at low temperatures most hydrogen prefers to collect at the various defects in the lattice.

Other key factors concern the diffusion of hydrogen to and from defects and the chemical reactions between hydrogen and inclusions. The absorption theory discussed depends on hydrogen being continuously available in the lattice to diffuse to defects, including the advancing crack. If stress is not important, what is the driving force that governs diffusion of hydrogen to these defects? In this respect, crack growth along grain boundaries appears to be the most plausible mechanism. Another area needing more work concerns the reactions that may occur between hydrogen and inclusions such as MnS, Fe_3C, silicates etc., (*see,* e.g., equation (4.15)). For example, it would be interesting to know the species of inclusions which react in this way and whether additional alloying affects these reactions. In any case, more work is needed to clarify some of these fundamental details before a satisfactory understanding of cold cracking can be developed.

Reheat cracking

There may be several reasons for reheating the weld metal and in multi-run welding several reheatings may be involved. However, the principal problem to be studied here concerns reheating the weld metal to temperatures in the range of *ca.* 500—650 °C in order to relieve residual stresses, and the particular form of cracking that may result from this treatment. In particular, certain types of steels, e.g. the austenitic stainless and ferritic creep resisting steels, and even the microalloyed steels, may be susceptible to reheat embrittlement. The phenomenon appears in general to be more associated with the HAZ (especially the grain growth zone) than the weld deposit.

Reheat cracking is thought to be closely related to the phenomenon of *creep rupture.* As is established in Chapter 3, the microstructure in the grain growth zone is likely to be relatively hard, particularly in alloy steels and high C_{equiv} steels. Furthermore, during reheating reprecipitation of carbides is likely to occur, especially on dislocations, which further increases the hardness. Because of this, creep deformation tends to be confined to grain boundaries and usually results in grain boundary sliding, *Figure 4.32.* In general the

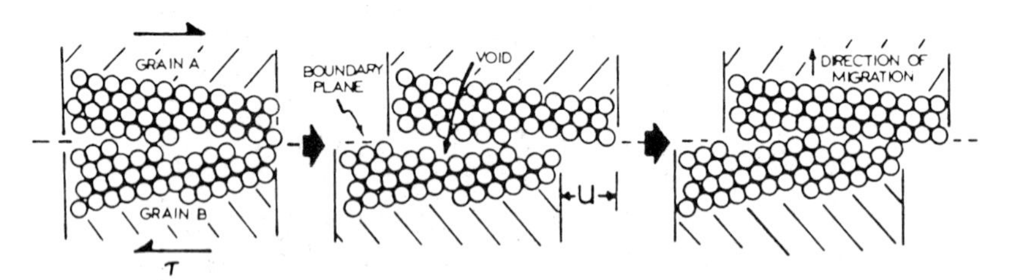

Figure 4.32 An atomic model of grain boundary sliding showing schematically how accommodation may occur. After Murr, L.E., *Interfacial Phenomena in Metals and Alloys,* Addison-Wesley, 1975

observed cracks are intergranular along the prior-austenite grain boundaries in transformable steels, and along any of the grain boundaries in austenitic steels.

Grain boundary sliding occurs more easily in small grained material since the strain distribution is more homogeneous in that case. In coarse grained material stress concentrations at the boundaries can cause them to rupture and it is this phenomenon which appears to occur in the grain growth zone during

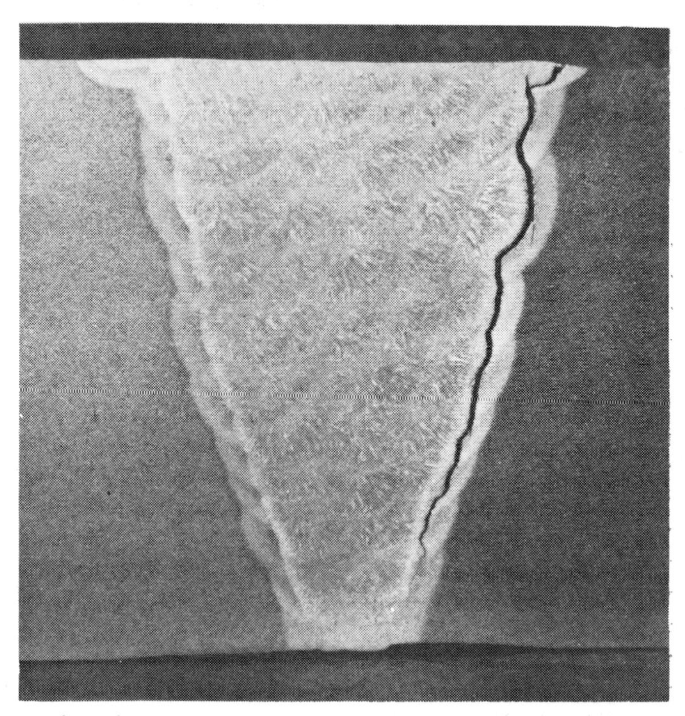

a) reheat crack in a CrMoV butt weld

x200

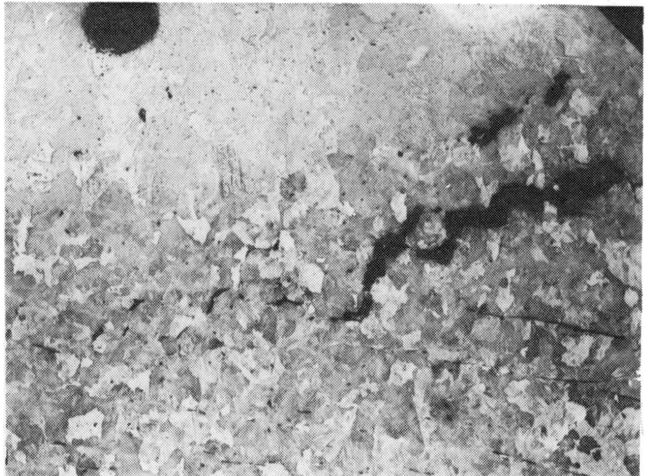

b) reheat crack in a structural steel weld

Figure 4.33 Examples of reheat cracks in the heat-affected zones of different steels. Note that the cracks are mainly intergranular. (a) is after Branah, G.D., High temperature mechanical properties as design parameters, *Rosenhain Centenary Conference*, p. 117, eds Baker, R.G. and Kelly, A., Royal Society, 1975; (b) is by courtesy of ESAB, Gothenburg (reduced by one-fifth in printing)

stress relieving. Indeed, in high strength metals, the creep strain required to nucleate cracks at boundaries need only be of the order of *ca.* 1 %.

Some examples of reheat cracks in welds are shown in *Figure 4.33*. The characteristic feature of these cracks is that they usually occur in patches rather than as a single path fissure, forming a relatively uniform, three-dimensional network in the grain growth zone. In this respect (*b*) in *Figure 4.33* is more typical of reheat cracking than (*a*). In some welds, cracks can also appear at coarse, columnar grain boundaries in the weld deposit. The form of these networks of cracks is dependent, to a certain extent, on the degree of restraint. In high restraint geometries in which triaxial stresses develop the cracks tend to occur in three-dimensional networks, but if uniaxial tensile stresses occur, the cracks will be unidirectional.

Reheat cracking is not confined to the transformable steels; the problem also occurs in the Ni-based alloys and austenitic stainless steels. Among the low alloy steels the most susceptible are the creep resistant grades of the type Cr–Mo–V, 2.5 Cr – 1 Mo and 1 Cr – 0.5 Mo[28]. As a general rule, however, post-weld heat treatment of the C–Mn steels improves fracture toughness[10], although low impact toughness has been reported for reheated microalloyed steels, particularly when welded by a high heat-input process[29].

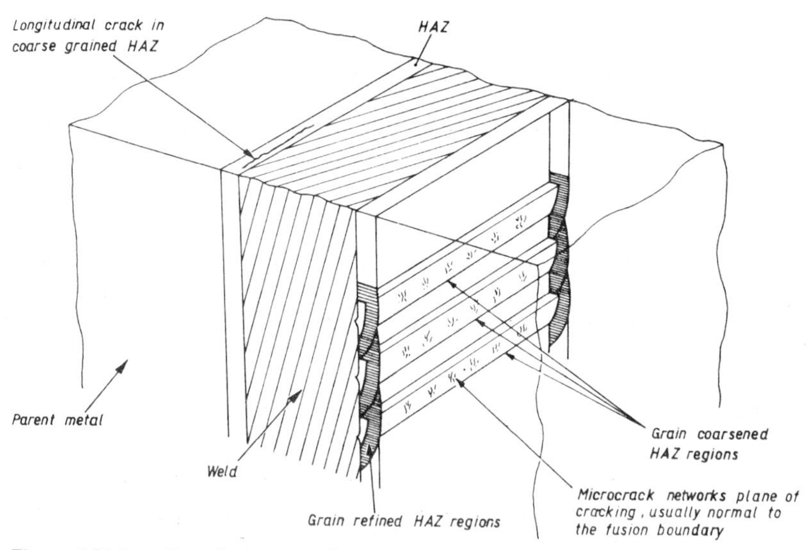

Figure 4.34 Location of reheat cracks in a nuclear pressure vessel steel. After Dolby and Saunders[30]

Of the alloy-steels, the 0.7 Ni – 0.4 Cr – 0.6 Mo steels used, e.g., in nuclear pressure vessels are well known to be susceptible to reheat cracking[30], as illustrated in *Figure 4.34*. Cracks occur in the grain growth zones either as longitudinal macrocracks, or (more often) as microcrack networks. This, of course, is a typical high restraint construction involving large residual stresses.

Design codes typically specify post-weld heat treatments of 550 – 650 °C for plate thicknesses in excess of 38 mm and for periods of an hour per 25 mm of thickness[10]. These specifications are based on isothermal relaxation tests in

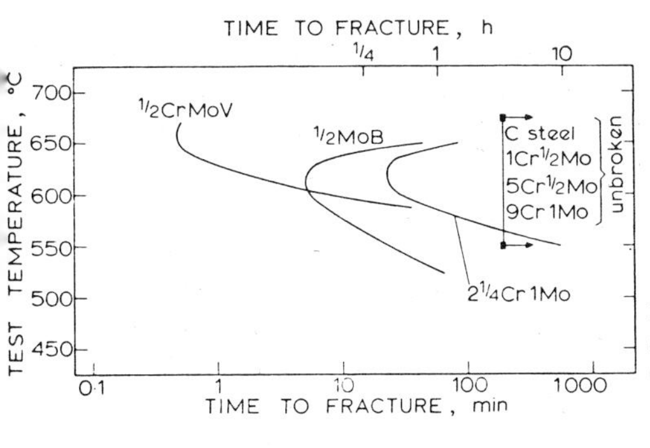

Figure 4.35 Notched stress-relaxation test results on HAZ-simulated creep-resisting steels. After Murray, J.D., Stress relief cracking in carbon and alloy steels, *British Welding Journal,* **14,** 447, 1967

which it is found that relatively short holding times at temperatures exceeding *ca.* 550 °C are sufficient for stress relieving of most types of steels. It is an unfortunate fact that this temperature range appears to closely correspond to (*a*) that in which reprecipitation of carbides readily occurs in the grain growth zone and (*b*) that where creep (at large grain sizes) occurs preferentially by a grain boundary sliding mechanism. The susceptibility of creep resistant steels to creep fracture in this temperature range is illustrated in *Figure 4.35.* It is instructive to study further these two key factors, particularly in the creep resistant steels.

Effect of reheating on the microstructure of a 0.5 Cr–Mo–V alloy

Perhaps the most researched creep resistant alloy, and its susceptibility to reheat cracking, is the Cr–Mo–V steel. Thus the microstructure of this alloy is considered herein, although many of the characteristics discussed apply equally well to other steels.

Cr–Mo–V steels are used, e.g., for turbine casings and pipework in high temperature heavy power equipment, because of their good creep resistance. They are, however, extremely prone to reheat cracking[31]. A typical composition (wt %) of these steels is[32]:

0.12 C; 0.62 Mn; 0.45 Cr; 0.54 Mo; 0.06 Nb; 0.25 V

with small traces of P, S, and other elements. The creep strength and stability of the steels is due to a refined microstructure with a fine dispersion of very stable carbides, including NbC, VC, $Cr_{23}C_6$, Mo_2C, etc. However, as seen from the carbide dissolution curves (*Figure 3.13*), during the weld thermal cycle these carbides probably dissolve in the austenite at temperatures above *ca.* 1200 °C, and grain growth can then occur virtually unrestricted. Using data of Alberry and Jones[32], a welding grain growth diagram of this alloy can be plotted[33] and this is shown in *Figure 4.36.* At the welding energies likely to be used in the MMA and SA processes, e.g. *ca.* 1–10 MJ m^{-1}, the microstructure consists of large grains, up to *ca.* 150 μm (compared to *ca.* 20 μm for the parent plate), which can extend for widths of up to *ca.* 10 mm from the fusion line. Cooling the weld produces a mixed bainite–ferrite microstructure, depending on the amount of preheat used (*see,* e.g., *Figure 3.33*), and some fine continuous

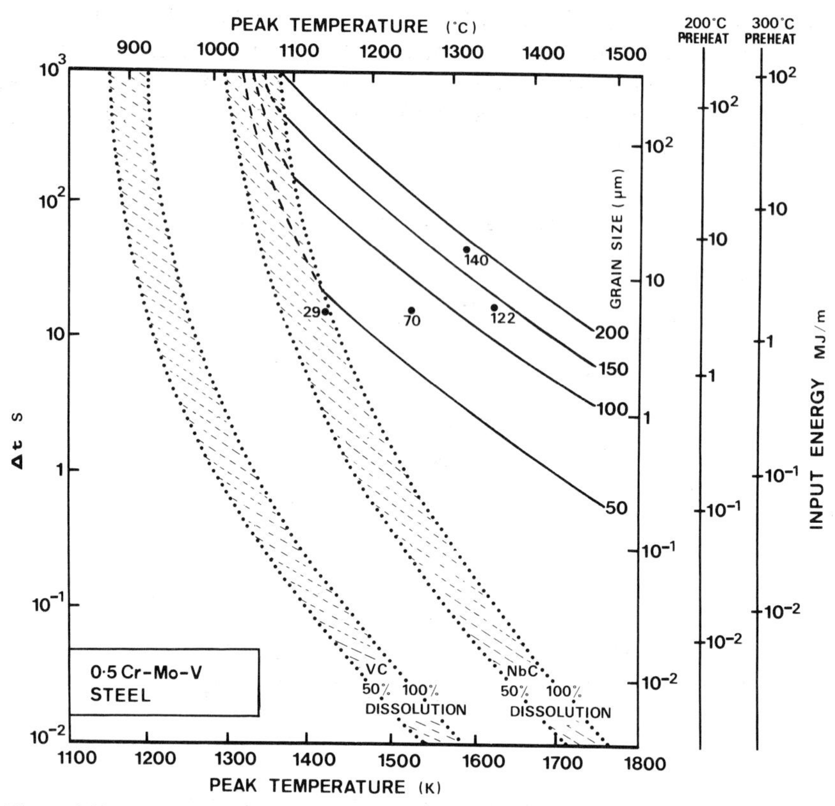

Figure 4.36 Austenite growth diagram for a 0.5 Cr–Mo–V alloy; T_0 = 20, 200 and 300 °C. The small figures refer to experimental measurements of austenite grain size, using a weld simulator. After Ashby and Easterling[33]

precipitation of VC[32]. The weld cycle is thus expected to produce a hard, coarse grained structure.

Reheating this alloy to *ca.* 600 °C brings about further precipitation of VC and some additional precipitates of NbC, which tend to form on dislocations[29]. The combined effect of cooling and reheating is thus to harden the grains even more. Reheating also promotes impurity segregation to grain boundaries[32], and this type of segregation is greater for the larger grain sizes.

The effect of grain size can be understood by comparing the ease with which impurities diffuse along a boundary of a given width, δ, with that through the lattice. Grain boundary diffusion becomes dominant under condition (4.21), where d is the grain size, and D_b and D_m are the grain boundary and

$$D_b\delta > D_m d \tag{4.21}$$

volume diffusion coefficients. Assuming that both types of diffusion act simultaneously, the effective diffusivity, D_{eff}, is given by equation (4.22). This

$$D_{eff} = D_m + D_b\left(\frac{\delta}{d}\right) \tag{4.22}$$

equation is plotted schematically in *Figure 4.37* which shows that, below a certain temperature, D_{eff} is dominated by grain boundary diffusion. The scaling factor (δ/d) for the case being considered would be *ca.* 3×10^{-3}, and it can be shown that grain boundary diffusion is likely to be dominant at temperatures below *ca.* 1000 °C.

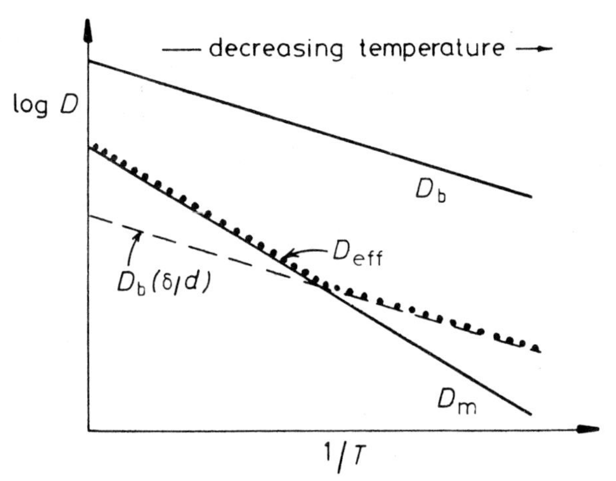

Figure 4.37 Grain boundary *vs.* lattice diffusion as a function of temperature in a metal. After Porter, D.A. and Easterling, K.E., *Phase Transformations in Metals and Alloys*, Van Nostrand Reinhold, 1981

The types of impurities moving to and along boundaries are largely determined by their grain boundary enrichment factor (*see* equation (3.36) and *Figure 3.27*). As seen in *Figure 3.27*, elements that have the highest segregation factors in α-Fe are (in order) sulphur, carbon and boron. Other elements commonly thought to segregate in Cr–Mo–V steels if they are present are P, Cu, Sn, As and Sb. The relative effect of these impurities on reheat cracking in

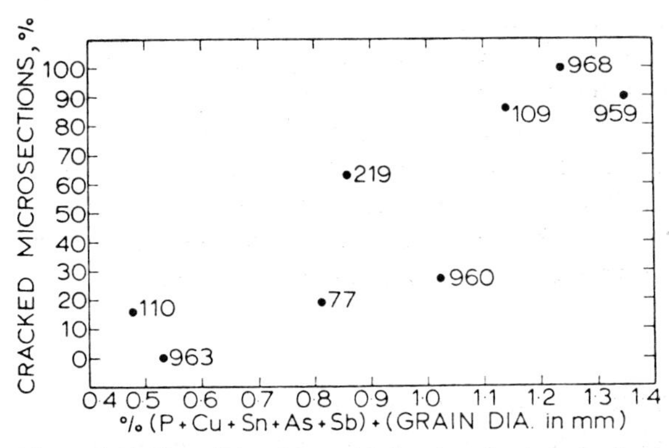

Figure 4.38 The effect of impurities and grain size on reheat cracking in various 0.5 Cr–Mo–V steels. The numbered experimental points refer to different specimens. After Boniszewski[28] (quoting results of R.L. Drinnan and P. Harris)

Cr–Mo–V steels as a function of grain size is shown in *Figure 4.38*. The grain size is thus important, both as a sink for impurities and also because of its effect on the mechanism of residual strain relief in welds.

Mechanism of reheat cracking

In the process of creep, the strain accommodation between grains and, particularly, at grain corners is normally accomplished by a combination of grain boundary sliding, vacancy diffusion and dislocation climb, *Figure 4.39 (a)*. The way grain boundary sliding and accommodation occurs on an atomic scale is

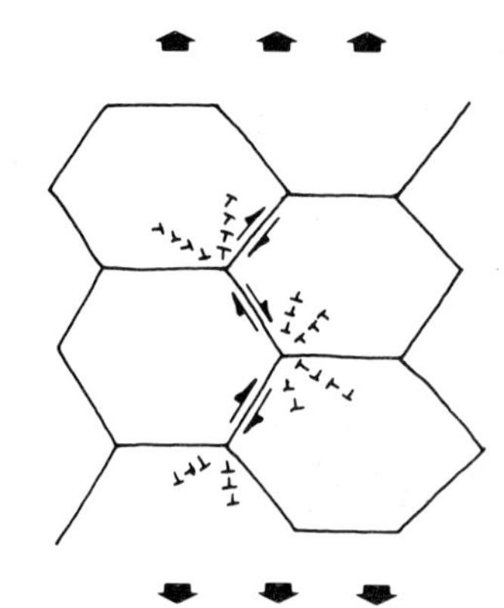

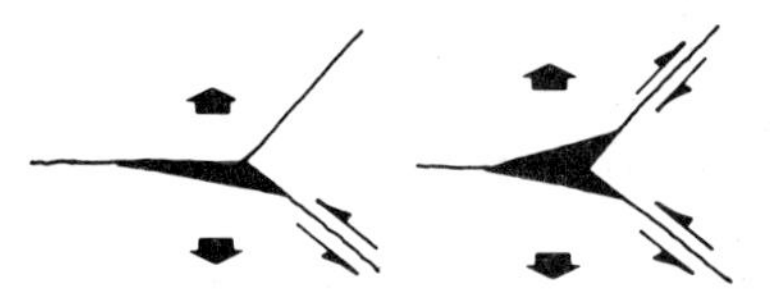

Figure 4.39 The accommodation of grain boundary (g.b.) sliding may occur by vacancy diffusion along the boundaries or by dislocation climb in adjacent grains, as shown in (a). If the grains are hardened and accommodation is hindered, cracks may initiate at junctions as shown in (b)

illustrated in *Figure 4.32*. The relative contributions to creep from volume (matrix) diffusion and grain boundary diffusion are a function of the stress, grain size and temperature. From equations (4.21) and (4.22), the total diffusional creep rate is given by equation (4.23)[34], where σ_a is the applied stress, Ω

$$\dot{\varepsilon} = \frac{2\sigma_a\Omega}{kT\,d^2} \times D_m\left(1 + \frac{2\delta}{d} \times \frac{D_b}{D_m}\right) \tag{4.23}$$

is the atomic volume, and k is Boltzmann's constant. As was shown earlier, the $(D_b\delta)$ factor dominates at temperatures below *ca.* 1000 °C. Furthermore, the average applied stress, σ_a, is likely to be well below the strength of the hardened grains. It is thus assumed that creep deformation is mainly restricted to grain boundaries.

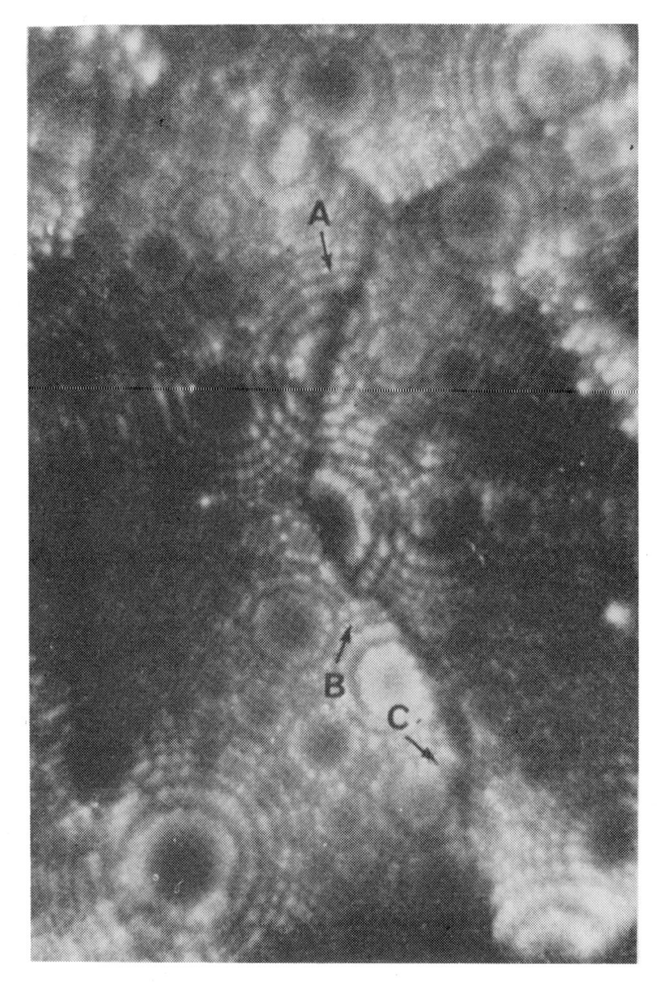

a) field ion micrograph showing g.b. ledges

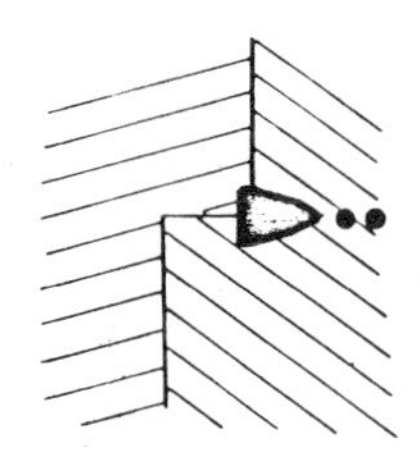

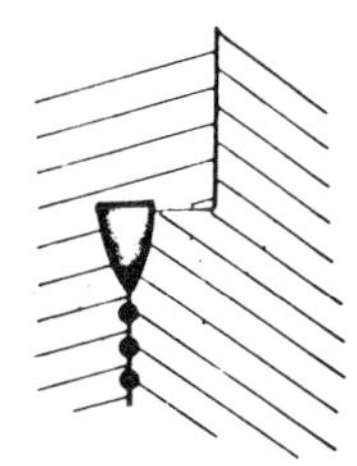

b) crack initiation at g.b. ledges

Figure 4.40 (a) A field ion micrograph showing the atomic structure of a grain boundary in tungsten, in which three ledges can be observed (marked A, B and C); after Ryan, H.F. and Suiter, J., Grain boundary topography in tungsten, *Philosophical Magazine*, **10,** 727, 1964. (b) Showing two possible mechanisms of grain boundary cracking at ledges; after Murr, L.E., *Interfacial Phenomena in Metals and Alloys*, Addison-Wesley, 1975

In metals that fail in a brittle, intergranular manner, fracture usually appears to originate within the boundaries themselves[35]. Cracks can either nucleate at inclusions and then propagate (as discussed, e.g., in connection with equations (4.18) and (4.19)) owing to reduced grain boundary cohesion because of impurity segregation or, if grain boundary sliding occurs, the cracks can nucleate at discontinuities in the boundary such as ledges or grain corners (triple junctions), *Figure 4.39* (*b*). A good example of ledges near a triple junction is shown in *Figure 4.40* (*a*). During grain boundary sliding this type of barrier can cause large stress concentrations. Furthermore, if, because of a hard matrix, plastic accommodation of these stress concentrations is unable to occur by slip, then microcracks tend to form in order to relieve the high stresses. The process of ledge-initiated fracture is illustrated in *Figure 4.40* (*b*). It seems likely that this mechanism of grain boundary rupture must also be of importance in reheat cracking, and be additive to the effect of impurity embrittlement.

Experimental studies of the complete process of intergranular creep fracture have revealed three sequential, but overlapping, stages[36]:

1. Nucleation of cavities on grain facets and triple junctions.
2. Growth of these individual cavities to form cracks of one grain facet in length.
3. Interlinkage of these single facet cracks to form cracks several grain diameters long which then rapidly lead to failure.

Concerning the effect of grain size on time-to-failure, a major fraction of the time can be spent linking up these single facet cracks to attain the critical crack

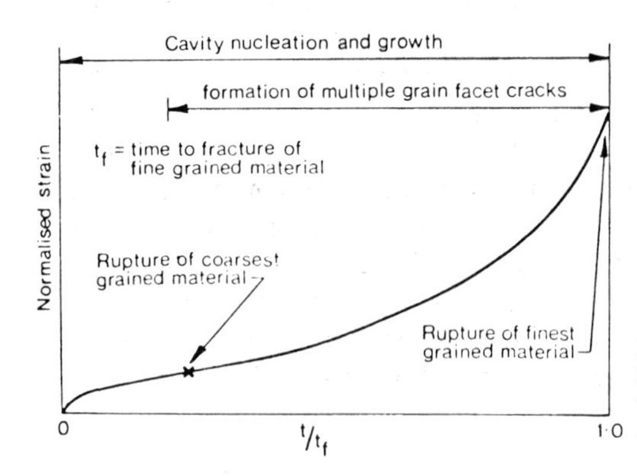

Figure 4.41 The differing importance of multiple grain facet cracks as a function of time to fracture, t_f, and grain size in Inconel. After Dyson anjd Taplin[36]

size needed for fracture. On this basis, it is found experimentally that coarse grained material fractures much faster than fine grained material, as shown schematically in *Figure 4.41*. This result again emphasizes the detrimental effect of grain growth in the HAZ.

It is common when using metals susceptible to reheat cracking to prescribe a preheating treatment in order to impose a slower quenching rate on the microstructure and thereby avoid an overhardened microstructure. It should be pointed out that a negative effect of preheating is that owing to the longer time spent at high temperature, grain growth and carbide solution can be increased substantially. This is demonstrated in *Figure 4.36* which shows that preheat effectively moves the energy-input to higher values. For example, it can be shown that in the case of a Cr–Mo–V steel, a 200 °C preheat can increase the maximum grain size in, e.g., MMA welds by *ca.* 50 μm. In view of the importance of grain size to creep rupture failure, this could have serious consequences. On this basis it is better to reduce the welding input-energy, if possible, when employing preheat.

It is thus concluded that, in general, the important factors that govern the cause, as well as the mechanism, of reheat cracking are broadly understood. The most important of these appear to be:

1. The development of *large grain sizes* in the HAZ during welding tends to increase segregation, to increase the likely number of ledges per boundary, and to reduce time-to-fracture.
2. The *reheating temperature* tends to encourage fine reprecipitation within grains and grain boundary segregation.
3. The *joint geometry* and *weld heat input* determine the amount of relaxation strain during reheating.
4. The presence of *impurities* can reduce grain boundary cohesive strengths.
5. The presence of *grain boundary particles,* may be detrimental or beneficial depending upon such factors as their size, interfacial energy, etc.
6. The use of *preheat* can substantially increase grain size.

Any improvements in reheat crack resistance have to be made by balancing these various factors against possible losses in creep strength of the alloys concerned.

In this chapter, the various factors that may contribute to the cracking and fracture of welds have been reviewed. In most cases, knowledge of the elements of fracture mechanics and physical metallurgy, as applied to welds, is sufficient to allow understanding of the various cracking phenomena and even enable certain predictions concerning susceptibility to cracking to be made. As an exercise in the application of these principles, the chapter is completed by considering a case study. As with most actual case studies, many factors have to be considered in analyzing a failure of this sort and the final conclusions have usually to be based on reasoned judgement rather than exact calculation.

Case Study: The *Alexander Kielland* disaster

On the evening of 27th March, 1980, a couple of minutes before 6.30 p.m., the *Alexander Kielland,* a drilling rig converted into an accommodation platform and located in the North Sea, started to capsize and within 20 minutes had overturned killing 123 of 212 people on board. The reason for the failure was later traced to a small 6 mm fillet weld which joined a non load-bearing flange

TABLE 4.3 List of Pentagone rigs

Name	*Owner*	*Country of registration*	*Ship yard*	*Delivery date*	*Area of operation*
Neptune 7	Forex Neptune and Associates	France	CFEM, Le Havre, France	June, 1969	North Sea, Mediterranean, West African coast
Pentagone 82	Forex Neptune and Associates	Panama	Marathon Le Tourneau Co., Texas, USA	October 1973	North Sea, Canada, Brazil
Drill Master	A/S Norsedrill and Co.	Norway	CFEM, Le Havre, France	November 1973	North Sea
Pentagone 84	Forex Neptune and Associates	Panama	Rauma Repola OY, Mantyluoto, Finland	October, 1974	North Sea
Venture One	Pel–Lyn Godager Co.	Panama	Rauma Repola OY, Pori, Finland	April, 1975	North Sea
Dixilyn–Field 96 (Venture Two)	Dixilyn–Field Drilling Co. (Pel–Lyn Godager Co.)	Panama	Rauma Repola OY, Pori, Finland	January 1975	North Sea
Dixilyn– Field 97 (Medusa)	Dixilyn–Field Drilling Co. *Offshore Drilling Inc.*	Liberia	France Offshore Group, *Le Havre, France*	January 1975	Mediterranean
Henrik Ibsen	Stavanger Drilling I	Norway	CFEM, Dunkirk, France	February, 1976	North Sea
Alexander L. Kielland	Stavanger Drilling II	Norway	CFEM, Dunkirk, France	July, 1976	North Sea
General Enrique Mosconi	Argentina	Argentina	CFEM, Le Havre, France	September, 1976	South Atlantic
Gulnare	K/S Morland Offshore A/S and Co.	Norway	CFEM, Dunkirk, France	January, 1977	North Sea, Mediterranean

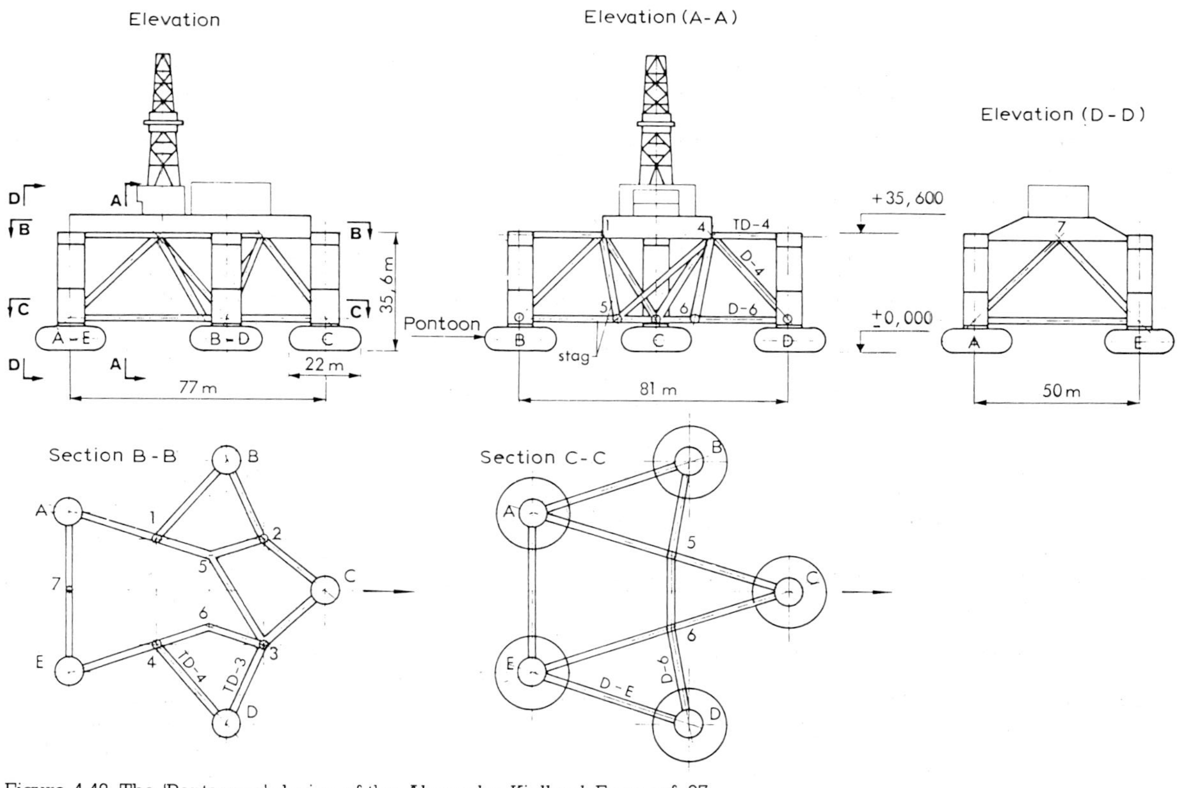

Figure 4.42 The 'Pentagone' design of the *Alexander Kielland*. From ref. 37

plate to one of the main bracings. The purpose of the flange plate was to hold a sonar device used in connection with drilling operations; ironically, the platform was never actually employed as a drilling rig. This case study is concerned with the possible factors that contributed to the failure of the weld. For further backgrounds detail, *see* ref. 37.

Construction of the *Alexander Kielland*

The *Alexander L. Kielland* was a mobile platform of the Pentagone type and was designed and built in France at the Dunkirk Shipping Yards. The rig was ordered by the Norwegians in 1973 and delivered on 5th July, 1976. It was originally built as a drilling rig, but during its entire operation it was utilized instead as an accommodation platform. Initially, its capacity was for 80 beds, although by April, 1978, this had been increased to 348. Altogether, eleven platforms of this type have been built, of which nine are currently in service in the North Sea, *Table 4.3*.

The characteristic form of a Pentagone design is shown in *Figure 4.42*. The main concern herein is the 'D' column and the bracing D-6. A detail of this part of the rig is given in *Figure 4.43* (*a*), which shows the location of the sonar flange plate. A detail of the flange plate as welded to the main bracing (D-6) is shown in *Figure 4.43* (*b*). The D-6 bracing is a *ca* 24 m long, circular, hollow beam of diameter 2.6 m and thickness 26 mm. It is left open to the sea and allowed to contain sea water in order to increase the rig's stability. For this purpose, the bracing contains an elongated opening (300 × 800 mm) located on the bottom of the bracing next to the sonar flange plate. Since no sonar device was ever fitted to the rig, this hole was also left open to the sea. Although not shown in *Figure 4.43* (*a*), an air hole of 150 mm diameter was located on the upper side of the bracing (this can actually be seen in *Figure 4.46* (*a*)). Both the air hole and the elongated opening were fitted with flanges in order to reduce the stress concentrations at these openings. This aspect is discussed later.

The production schedule at the Dunkirk Yards was such that the assembly work was divided between two teams, one team being responsible for the main welding and fitting operations and the other taking care of the welding and fitting of ancillary equipment. In this respect, for example, the welding of the flanges to the elongated opening and air hole was included among the duties of the main installation team, while the welding of the non load-bearing sonar flange plate was the responsibility of the other team. Furthermore, it was not considered necessary in the design work to carry out any stress analysis of the sonar flange plate fitting, although a stress analysis of, e.g., the oval hole flange plate was carried out. This turned out to be a vital omission. The main braces were of a welded construction and made from a Nb-microalloyed fine-grained steel.

The construction and fitting of the sonar flange plate

As shown in *Figure 4.43* (*b*), the flange plate is essentially a short, circular, hollow cylinder, *ca.* 228 mm long and 325 mm diameter, with wall thickness 20 mm. Similar flange plates were fitted to three of the main braces, i.e. B-5,

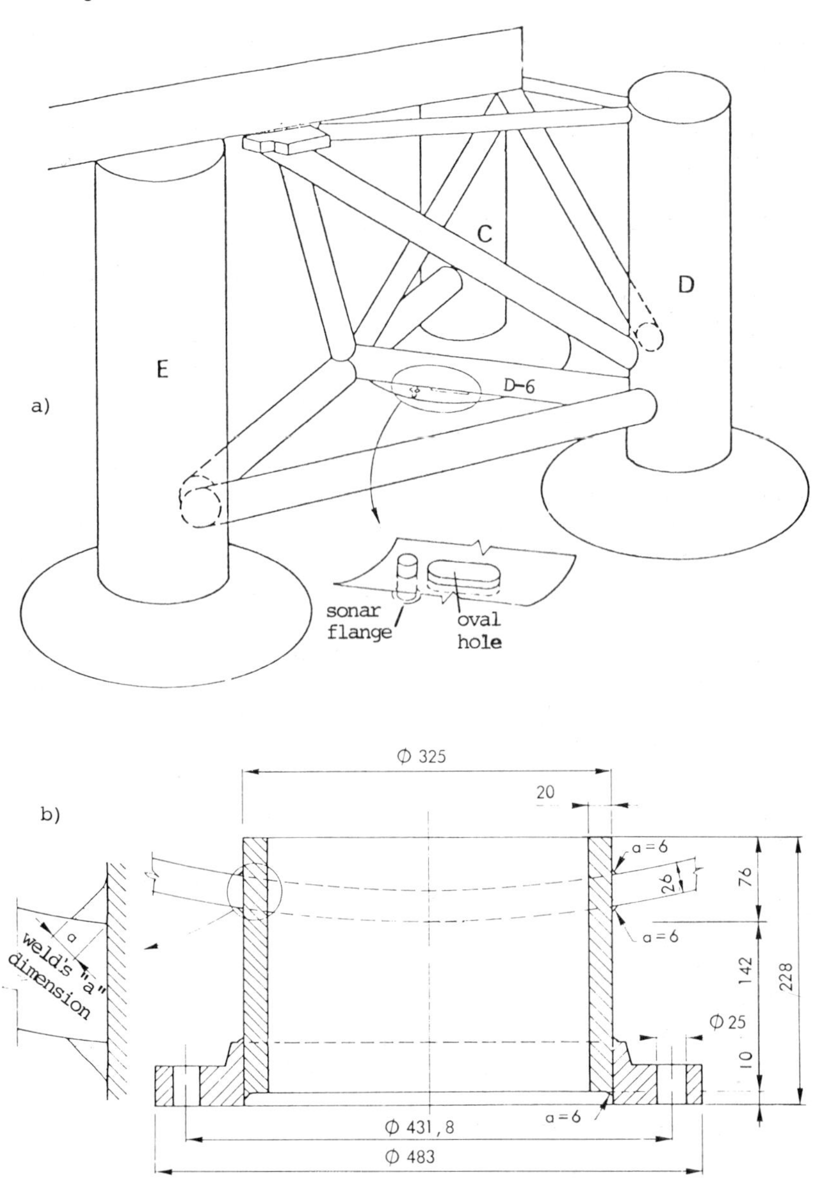

Figure 4.43 (a) The location of the 'D-6' bracing, showing the position (insert) of the sonar flange and oval hole. (b) A detail of the sonar flange plate; the insert defines the 'a-dimension' of the weld. After ref. 37

D-6 and A-5. The flange plate material was of a fine-grained pearlitic–ferritic steel, shaped by bending and butt welding. The profile of the butt weld was of an 'X' form, i.e. it was welded both from inside and outside employing 2 runs on the inside and up to four on the outside. The welding method was MMA, using flux-covered electrodes.

The sonar flange plate was located at a flame-cut hole in the bracing, of approximately 3—5 mm larger diameter than the flange itself. The flange plate

Figure 4.44 The *Alexander Kielland* photographed a few weeks before the disaster. From ref. 37

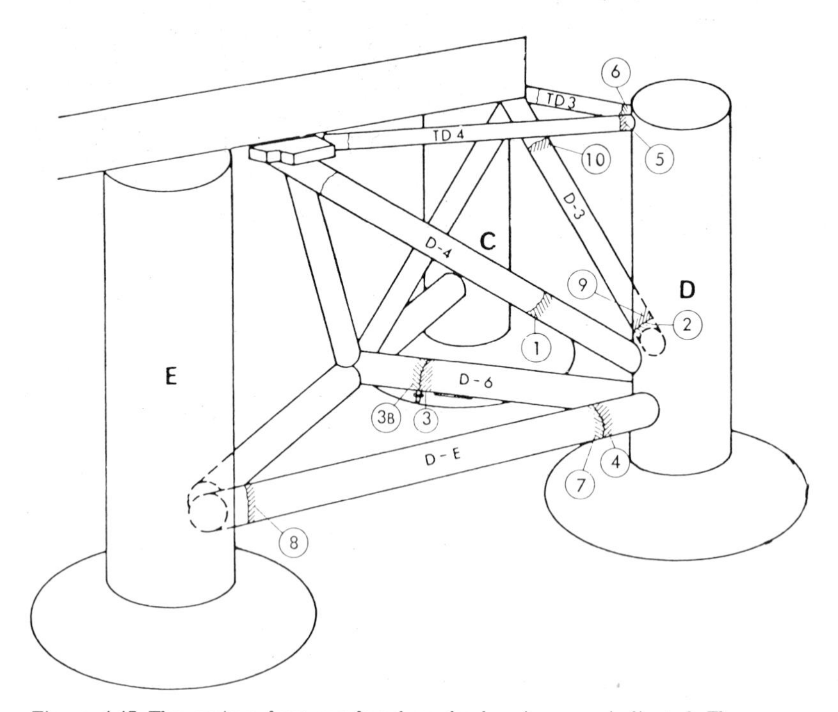

Figure 4.45 The various fractures found on the bracings are indicated. The cross-hatching and numbers refer to sections investigated metallographically (ref. 37)

was then welded in position using MMA welding of 2–3 runs per weld, employing fillet welds both inside and outside the main brace plate, *Figure 4.43 (b)*. Flux-covered electrodes of 'basic' type, 5 mm diameter, were specified for this purpose. The '*a*'-dimension of the weld was given as 6 mm, but the number of runs per weld was not specified. Preheat was neither specified nor employed.

Capsize of the *Alexander Kielland*

On the 27th March, 1980, the day of the disaster, the *Alexander Kielland* lay at anchor in the Ekofisk field close to the production platform *Edda 2/7 c*. A view of the *Alexander Kielland* shortly before the disaster is shown in *Figure 4.44*. The weather in the North Sea on that fateful day was stormy with mist and

(a)

(b)

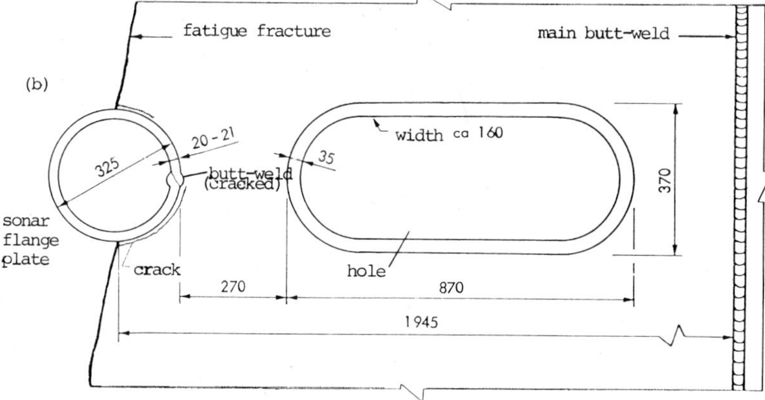

Figure 4.46 (a) The recovered bracing D-6. (b) A detail of the fracture position relative to the sonar flange plate. After ref. 37

rain and visibility down to about a kilometre. It was also cold, with an air temperature of 4—6 °C and a sea temperature of 6 °C.

As the day progressed the weather deteriorated, with the wind blowing at 20 m s^{-1}, churning up waves of 6–8 m in height. With the storm worsening, it was decided, at about 6.00 p.m., to move the *Alexander Kielland* away from *Edda 2/7 c*, and the gangway connecting the two was hoisted on board the *Kielland*.

About half an hour later, at 6.28 p.m., the radio officer on board the *Kielland* heard a loud thump from below. Not too much notice was paid to

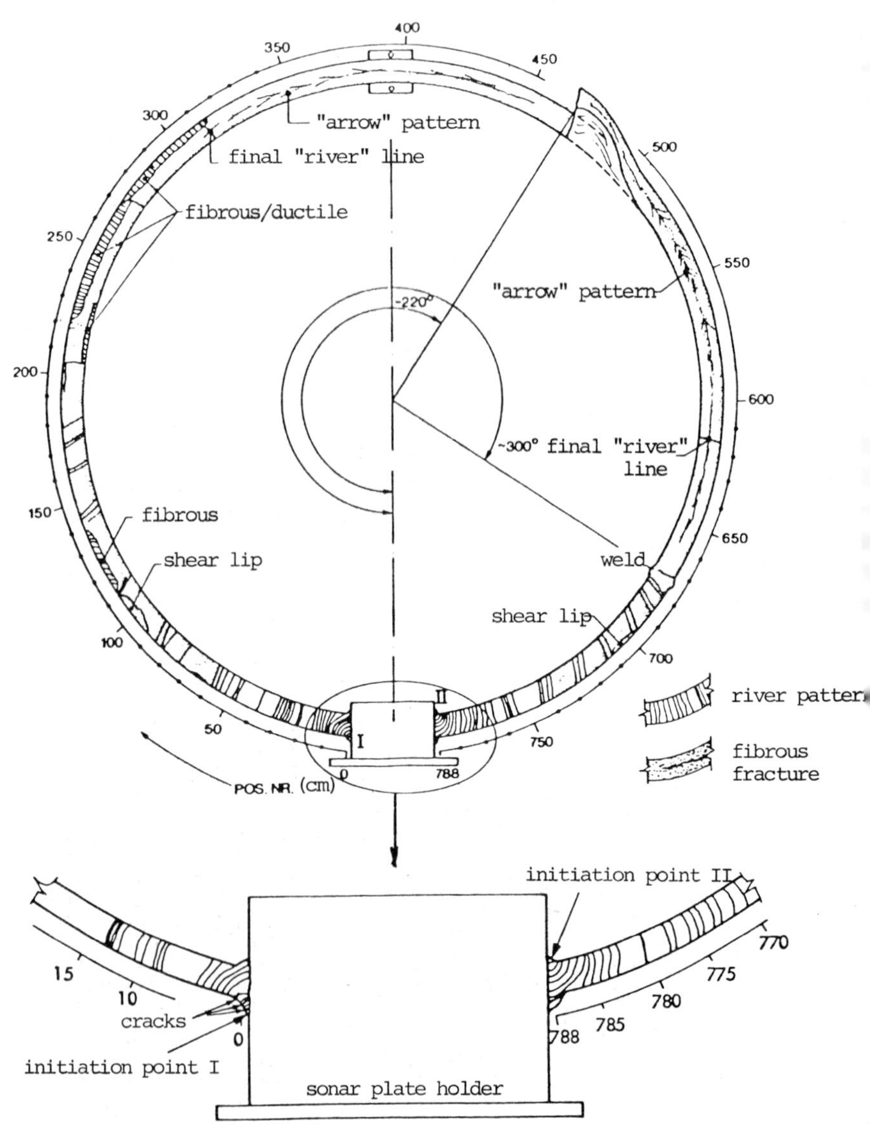

Figure 4.47 Mapping of the main fracture surface as based on metallographical investigation. The detail below shows that the crack initiated on either side of the sonar flange plate. After ref. 37

begin with, since such noises are not unusual in heavy seas. Soon after the first thump, however, came another and this was followed by a definite listing of the platform. The radio officer immediately sent out a 'Mayday' call which was picked up by the *Edda* at 6.29 p.m. Minutes after the second thump was heard, the platform had already listed over to an angle of 30–35° from the horizontal. Indeed, from *Edda* it was apparent that only the anchor wire, 'as taught as a violin string', prevented the platform from turning over completely. At 6.53 p.m., 24 minutes after the 'Mayday' was sent out, it was recorded in the log on board the *Edda* that the *Alexander Kielland* had completely overturned.

It was later established by the Norwegian Commission that investigated this incident (ref. 37), that the first thump heard by the radio officer was certainly caused by the fracture of the main brace, D-6. Then followed, in rapid succession, failures of the other bracings which connected column D to to the platform, these resulting presumably from overloading. The positions of the various fractures of the bracings are shown in *Figure 4.45*. The spacing of the latter fractures led the Commission to conclude that failure of bracings other than D-6 was due to bending.

The failure in bracing D-6, which initiated the structural failure, was clearly due to fatigue. Indeed, it was later established that prior to the final fracture, the crack had grown to a length of over 5 m, or *ca.* 2/3 the circumference of the bracing! *Figure 4.46* (*a*) shows the recovered D-6 bracing with the positions of the elongated hole and sonar flange plate indicated. *Figure 4.46* (*b*) is a detail from the flange plate region which shows that the main fracture had occurred quite independently of the main butt weld of the main bracing (shown to the right); it is, however, clearly associated with the fillet weld of the flange plate. Studies of the characteristic river patterns of the main fracture confirm this, as illustrated in *Figure 4.47*, in which the fracture pattern has been mapped out. As shown in the detail of *Figure 4.47*, fatigue initiated at two parts of the fillet weld, first at point I and then at point II.

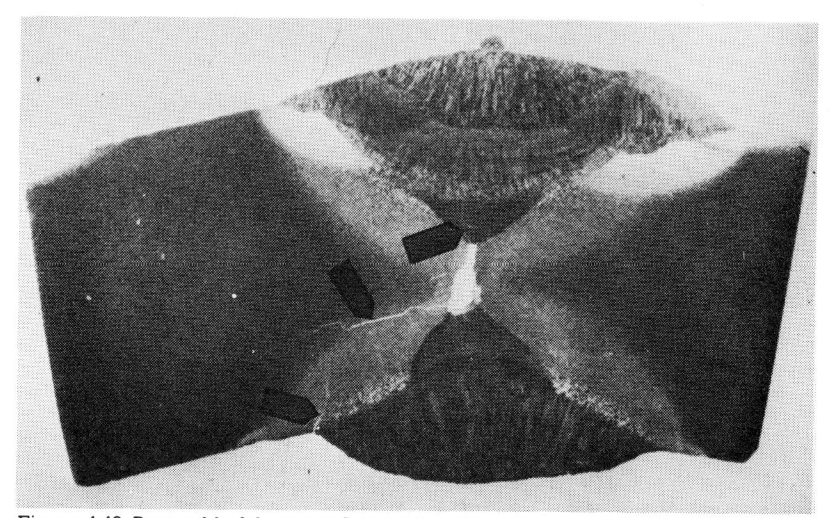

Figure 4.48 Butt weld of the sonar flange plate. Root and toe cracks, as well as lamellar tearing are indicated by markers. After ref. 37

Metallographic examination of the sonar flange plate welds

A metallographic examination of the fractured D-6 bracing revealed (amongst other things) the following factors[37]:

1. The *butt weld* of the sonar flange plate contained both toe and root cracks, the latter extending the whole length of the weld. Lamellar tearing in the flange plate material was also observed. Some of the cracks are illustrated in the micrograph of *Figure 4.48*.
2. Secondary cracking associated with the butt weld was observed at the cross-over between the butt weld and the fillet weld.
3. The quality of the butt weld was generally poor, exhibiting unsatisfactory penetration of the base material.

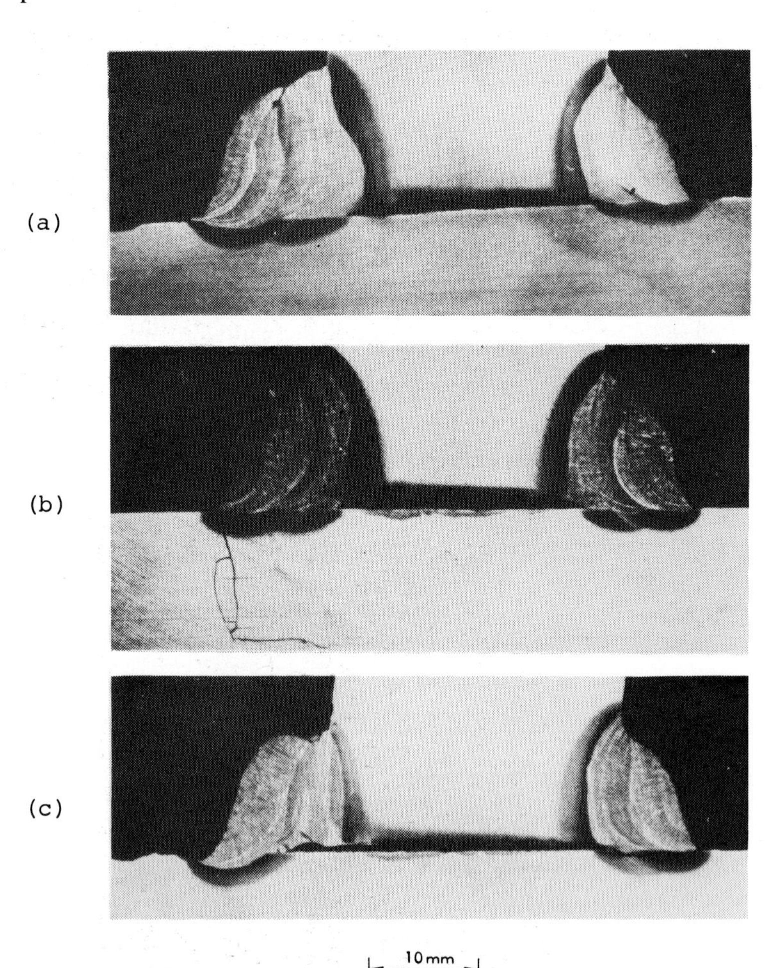

Figure 4.49 Three different sections of the fillet weld between the flange plate (horizontal) and main bracing (vertical). Note the possible toe cracking and lamellar tearing of the flange plate in (b), and the poor penetration of the weld shown in (c). After ref. 37

4. The quality of the *fillet welds,* connecting the flange plate to the main bracing, was generally poor, showing insufficient penetration and uneven profiles, *Figure 4.49*. Indeed, the '*a*'-dimension (specified as 6 mm) was found to vary in practice between 5 and 9 mm.
5. Significant amounts of lamellar tearing of the flange plate material were observed in association with the fillet welds. However, there was no evidence of lamellar tearing in the bracing plate material.
6. Cracks running parallel to the edges of the fillet welds and joining up with the main fracture were observed (*see,* e.g., *Figure 4.46* (*b*)).
7. Several small cracks were found associated with the fatigue initiation point I, as illustrated in *Figure 4.50*. The appearance of these cracks is not unlike that associated with cold cracking (*see,* e.g., *Figure 4.18*).

(a)

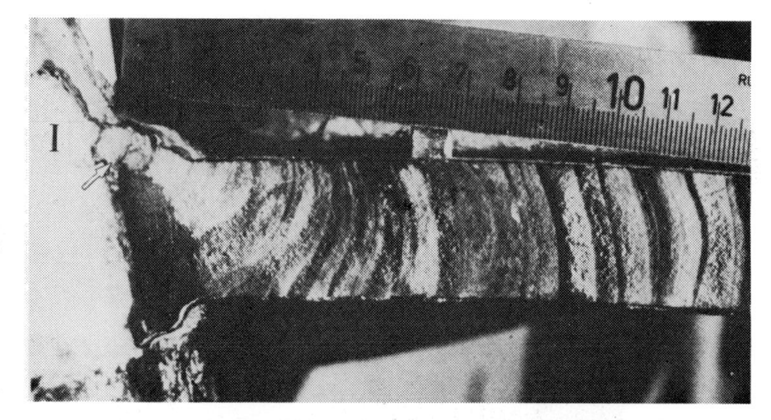

(b)

Figure 4.50 Two details of the main initiation point of the crack in the fillet weld between the bracing (horizontal plate) and the sonar flange (vertical). The river pattern of the fatigue crack is clearly visible in (a). Several fine cracks are seen in the weld in (b). After ref. 37

8. Small traces of paint, of the type originally used in the Dunkirk Yards, were discovered in the fracture surface of the fillet weld, implying that some cracking had occurred in this weld already during manufacture of the rig and prior to it going to sea.
9. Macrohardness measurements were carried out on the fillet-welded joint, and these results are illustrated in *Figure 4.51*. Significantly, the maximum hardness is associated with the HAZ, reaching values of *ca.* 350 HV_5 compared with *ca.* 160 in the base material. The hardness of the as-solidified weld metal lies between these values.
10. Mechanical property measurements of both the flange plate and bracing plate materials were carried out and these results, together with micro-

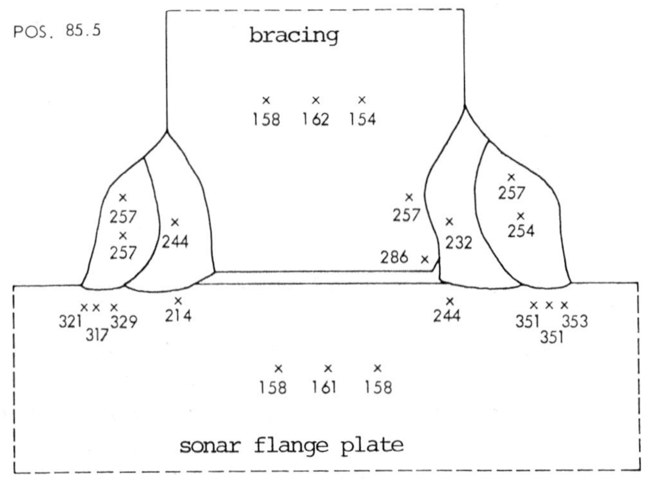

Figure 4.51 Hardness measurements (HV_5) in the vicinity of the fillet weld between the flange plate and main bracing (ref. 37)

TABLE 4.4 Summary of microstructural and mechanical properties

Property	*Main bracing*	*Flange plate*
Composition (approx.)	0.17wt%C; 0.32 Si; 1.37 Mn; 0.044 Al; 0.029 Nb	Not specified
C_{equiv}	0.41	Not specified
Grain size (normalized)	ASTM 11.5 (*ca.* 8 μm)	ASTM 10.7 (*ca.* 7 μm)
Microstructural features	1. Banded, ferritic–pearlitic 2. Slag content low; mainly MnS, finely distributed	1. Banded, ferritic–pearlitic 2. Slag content fairly high; mainly MnS in extreme rolled-out form
σ_y (rolling direction)	345–353 N mm^{-2}	Not specified
σ_t (rolling direction)	506–518 N mm^{-2}	Not specified
σ_t (transverse direction)	419–474 N mm^{-2}	215–437 N mm^{-2}
Area reduction (rolling direction)	30–34%	Not specified
Area reduction (transverse)	6–13%	1–7%
Impact strength (Charpy)		
0 °C	83 J	Not specified
−40 °C	36 J	16 J

structural observations, are summarized in *Table 4.4*. Note that although the grain sizes in both materials are small, the transverse (through-thickness) properties of the flange plate material are exceptionally poor. Fatigue life experiments were also carried out on the main bracing material, giving satisfactory values for the stress ranges expected (*see* ref. 37).

11. Stress analysis of the effect of the sonar flange hole on stress concentration in the bracing material, with and without the flange plate in position, was carried out. This showed that if the flange plate had not been present the stress concentration would almost double, reaching a value of about three times that of the mean stress in the bracing. If the flange plate had been intact, the maximum stress concentration would be about 1.8.
12. Examination of bracing A-5 and B-5, also containing sonar flanges, revealed no obvious signs of failure.

Possible effects of the weld thermal cycle on the bracing and flange plate materials

The bracing material The bracing material was a Nb-microalloyed, fine grained, high strength steel, *Table 4.4*. In normalized conditions the steel contains an extremely fine dispersion of NbC precipitate which stabilizes and refines the grain size. The carbon content is relatively high (0.17 wt %), giving the steel its fairly high C_{equiv} value (0.41). Aluminium (0.044 wt %) is mainly present as a deoxidizer, there evidently being little or no nitrogen present to form AlN or Nb(CN). The continuous cooling diagram for the steel should not be unlike that shown in *Figure 2.44* (for the 0.19 wt % C boundaries), in which it is seen that the relatively high cooling rates of MMA welding are likely to give a mainly bainitic microstructure. The presence of Nb, if it is in solution in the austenite, further tends to promote the lower bainite transformation, as indicated in *Figure 2.50*. In other words, as discussed on pp. 105–112, this alloy is a typical example of today's sophisticated steels which aim to possess the ideal combination of high strength and acceptable weldability.

As indicated in *Table 3.1* (p. 115), the solubility of NbC in steel is given by equation (4.24). For the steel in question the *equilibrium* solution temperature

$$\log_{10} [\text{wt \% Nb}]\,[\text{wt \% C}] = 2.96 - \frac{7510}{T} \tag{4.24}$$

is thus estimated to be *ca.* 1150 °C. During a rapid thermal cycle, as in welding, the carbide is effectively superheated, as shown in *Figure 3.13* (p. 122). It is to be noted that the NbC content in the bracing material is somewhat higher than that assumed in computing the NbC curve by *Figure 3.13*. However, the *shape* of the NbC curve in *Figure 3.13* is little affected by changes in NbC composition, it being mainly dependent upon the activation energy constant for the diffusion of Nb in austenite, equation (4.25). On this basis, knowing the welding energy used

$$\frac{Q}{RT_m} = 23.2 \text{ (for Nb)} \tag{4.25}$$

when carrying out the fillet weld, it may be estimated whether or not the NbC is likely to have gone into solution. According to the data available[37], the fillet weld was carried out with a 5 mm covered (basic) electrode of the type OK 48.30, manufactured by ESAB. The ESAB brochure gives the following data for this electrode:

Current: 200–260 A
Voltage: 24 V

Assuming an efficiency for MMA welding of 0.8 (*Table 1.4*), an input energy of about 0.7—0.8 MJ m^{-1} can be assumed. This corresponds to a Δt_{8-5} of about 2—3 s, giving a temperature of solution for NbC of *ca.* 1270 °C.

The implications of this are threefold. In the first place, it means that grain growth in the HAZ at temperatures above *ca.* 1270 °C can occur unimpeded. In addition, it is likely that Nb in solution tends to promote the lower transformation products on cooling. Finally, the very rapid cooling rate through

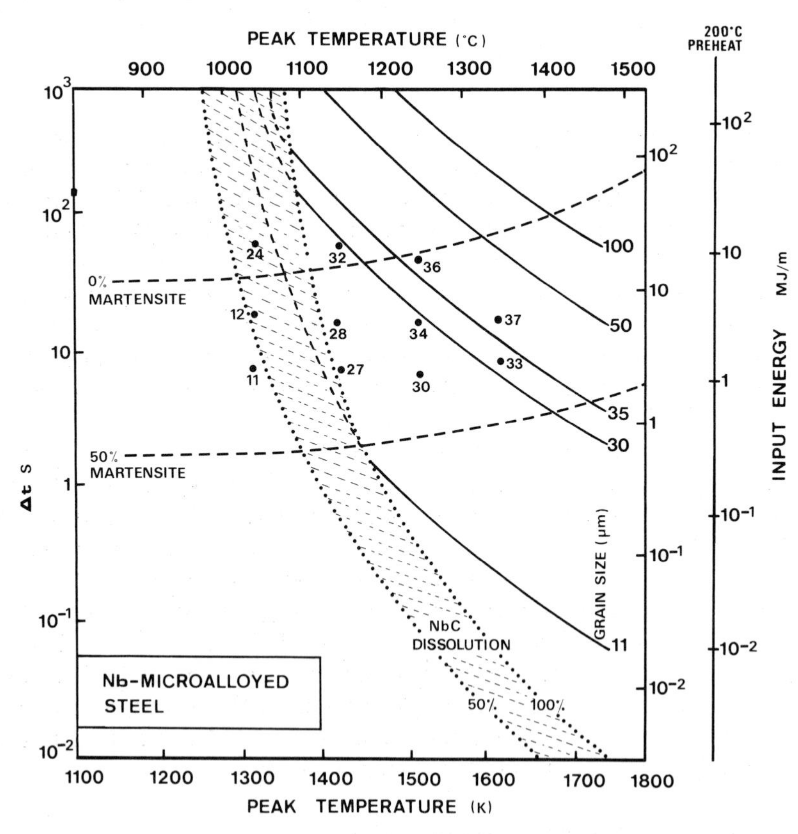

Figure 4.52 Austenite grain growth diagram for a bead-on-plate welded Nb-microalloyed fine grained steel for T_0 = 20 and 200 °C (thick plate condition). The small figures in the diagram refer to experimental measurements of austenite grain size in simulated samples. The dotted curves refer to amounts of martensite as measured in the simulated samples. After Ashby and Easterling[33]

800→500 °C predicted for this weld (2—3 s), tends to promote the martensite transformation in this steel.

On the basis of considerations developed on pp. 126–130, it can be estimated that the maximum austenite grain size for this weld energy-input (assuming a thick-plate condition) is of the order of *ca.* 30–35 μm (*Figure 4.52*), compared with *ca.* 8 μm for the ferritic base material.

Thus, the Nb in solution in the region of the HAZ and the high cooling rate are both likely to produce a hard bainitic–martensitic microstructure in the HAZ of the bracing material. The conclusion is supported to a certain extent by the hardness measurements (*Figure 4.51*) carried out on the fillet weld in question. It is unfortunate that no microhardness measurements or detailed microstructural studies of the HAZ of this weld were reported.

In the light of these conclusions, the presence of a number of small, underbead cracks in the vicinity of the HAZ of the fillet weld with the bracing material (*Figure 4.50*) is intriguing. The use of a basic electrode should normally keep the hydrogen content at a low value. On the other hand, it has been well established that the stress concentration at the weld was likely to be high and the microstructure is likely to contain martensite. On this basis, cold cracking of the fillet weld cannot be ruled out, even in the presence of low hydrogen levels.

The flange plate material Unfortunately, no compositional analysis for this material has been given in reference 37. Micrographs show, however, that the steel contains a highly banded, ferritic–pearlitic microstructure of a somewhat finer grain size than that of the bracing material (*see Table 4.4*). On the other hand, the MnS-inclusion content is not only higher than for the bracing material, but is present in the undesirable shape of long, rolled out laths. This, of course, gives the material its very poor transverse ductility (1–7 %). It is therefore hardly surprising that significant lamellar tearing was observed in connection with both the fillet and butt welds, *see,* e.g., *Figure 4.49* (*a*) and (*c*), and *Figure 4.47.*

The root cracks and toe cracks reported for the butt weld (*Figure 4.48*) are presumably due to cold (hydrogen) cracking. Assuming the use of the same basic electrodes as specified for the fillet weld, cold cracking problems should not normally occur, although it is not known whether the electrodes were properly dried prior to use. Another possible contributary cause to these cracks may be in the form of liquation cracking, this often arising if elongated MnS particles intersect the fusion zone, as seems likely in the case of the butt weld (*see* pp. 170–172). In any case, the hardness measurements in the HAZ of this material (*Figure 4.51*) are somewhat higher than for the bracing material HAZ. This suggests the presence of a significant amount of the bainitic–martensitic structure, known to be susceptible to cold cracking.

Mechanism of failure: main conclusions

There appears to be a number of factors which could possibly have contributed to the failure of the D-6 bracing, with the resulting capsize of the *Alexander Kielland.* Of these the following are probably the most important:

1. The considerable amounts of lamellar tearing in the flange plate and the extensive root crack in the butt weld both contribute to the weakening of the sonar flange-plate in its capacity as structural strengthener. This, together with the resulting increased stress concentration at the hole, evidently induced cracking (or caused existing cracks to grow) around the periphery of the flange plate in the fillet weld.
2. The poor profile of the fillet weld contributed to a reduction in fatigue strength of the weld.
3. The rapid cooling rate of the fillet weld, the dissolution of NbC precipitates, and some grain growth in the HAZ of the bracing plate, all (together with the increased stress concentrations) helped produce just the conditions likely to give rise to cold cracking.
4. Given the presence of cold cracks in the fillet welds, the increase in stress concentration due to weakening of the flange plate, the poor weld profile, and the cyclic stresses experienced at sea then all the necessary conditions for fatigue crack growth appear to be present.

As with most case studies, there are some unanswered questions. For example, why did significant cracking occur in bracing D-6, but not in bracings A-5 and B-5, which also contained sonar flange plates? A possible variable might be the dryness of the electrodes used for the fillet weld, particularly in view of the fact that this weld was considered of secondary importance at the manufacturing stage. It is also not clear that the loading conditions are the same for all three bracings under normal service at sea, and this may be another important variable.

Yet another factor of interest is that the main welds of the bracing were evidently not affected by cold cracking. However, there are at least two possible reasons for this. In the first place, at the manufacturing stage these welds were considered of primary importance and no doubt received additional care when made. In addition, these larger (possibly automatic) welding operations imply the use of higher input-energies with consequently longer Δt_{8-5} cooling times, and this would result in a lower martensite content in the HAZ in spite of the larger grain size (*see,* e.g., *Figure 4.52*).

In spite of the uncertainties involved, this case study emphasizes the important role that can be played by physical metallurgy in improving prediction of weldability and analysis of welding problems. Perhaps too much emphasis in the past has been placed on empirical methods for predicting weldability, although the complexities involved are many and this approach may have been justified. Nevertheless, the challenge offered to future physical metallurgists in this respect is formidable and exciting.

References

1. Richards, K.G., *Brittle Fracture of Welded Structures,* The Welding Institute, Abington, 1971

2. Barr, R.R. and Burdekin, F.M., Design against brittle failure, *Rosenhain Centenary Conference Proceedings,* p. 85, eds Baker, R.G. and Kelly, A., Royal Society, 1975

3. Reed-Hill, R.E., *Physical Metallurgy Principles,* Van Nostrand Reinhold, 1964

4. Harrison, J.D., Davies, M.G., Archer, G.L. and Kamath, M.S., *The COD Approach and its Application to Welded Structures,* The Welding Institute, Report No 55/1978/E, Abington, 1978

5. Pickering, F.B., Sheffield City Polytechnic, personal communication, 1982

6. Morgan-Warren, E.J. and Jordan, M.F., A quantitative study of the effect of composition on weld solidification cracking in low alloy steels, *Metals Technology,* **1,** 271, 1974

7. Nakata, K., Miyanaga, Y., Matsuda, F., Tsukamoto, K. and Arai, K., New Al–7 % Mg welding electrode, for crackless welding of Al–Zn–Mn (7NOI) high strength aluminium alloy, *Transactions JWRI,* **9,** 63, 1980

8. Roberts, W., Lehtinen, B. and Easterling, K.E., SEM observations of the development of cavities around inclusions in steels during plastic deformation, *Acta Metallurgica,* **24,** 745, 1976

9. Garland, J.G., *Solidification Cracking during Submerged-arc Welding of C–Mn Steels–the Relative Importance of Parent Strength and Composition,* The Welding Institute, Report No 25/1976/M, Abington, 1976

10. *Guide to the Welding and Weldability of C–Mn Steels and C–Mn Microalloyed Steels,* International Institute of Welding, Publications Document No IIS/IIW–382–71, 1971

11. Ashby M.F., Work hardening of dispersion hardened crystals, *Philosophical Magazine,* **14,** 1157, 1966

12. Farrar, J.C.M., Charles, J.A. and Dolby, R.E., Effect of second-phase particles on the mechanical properties of steel, p. 171, *Conference Proceedings of the Iron and Steel Institute,* London, 1971

13. Complementary information test for lamellar tearing, *Welding in the World,* **19**, 47, 1981

14. Fikkers, A.T. and Muller, T., Hydrogen induced cracking in weld metal, *Welding in the World,* **143,** 238, 1976

15. Fast, J.D., *Gases in Metals*, Philips Technical Library, 1976

16. Beck, W., O'M Bockris, J., McBreen, J. and Nanis, L., Hydrogen permeation in metals as a function of stress, temperature and dissolved hydrogen concentration, *Proceedings of the Royal Society,* **A290,** 220, 1966

17. Hart, R.H.M. and Watkinson, F., *The Implant Test as a Means of Ranking the Resistance to Hydrogen Cracking of Cr–Mo Weld Metals,* The Welding Institute, Report No 32/5/75, Abington, 1975

18. Andersson, B., *Hydrogen Cracking in Weldments,* Ph.D. thesis, Chalmers University of Technology, Gothenburg, 1981

19. Ochiai, S.I. and Kikuta, Y., Hydrogen delayed cracking of steels and its prevention assessment using computer simulation of stress (strain) induced behaviour, *Transactions JIM,* **21,** 473, 1980

20. Thompson, A.W., The behaviour of sensitized 309S strainless steel in hydrogen, *Material Science and Engineering,* **14,** 253, 1974

21. Garber, R., Bernstein, I.M. and Thompson, A.W., Effect of hydrogen on ductile fracture of spheroidized steel, *Scripta Metallurgica,* **10,** 341, 1976.

22. French, J.E., Weinrich, P.F. and Weaver, C.W., Hydrogen embrittlement of spheroidal steel, *Scripta Metallurgica,* **13,** 285, 1979

23. Cialone, H. and Asaro, D.J., Hydrogen assisted fracture of spheroidized plain carbon steel, *Metall Transactions,* **12A,** 1373, 1981

24. Porter, D.A., Easterling, K.E. and Smith G.D.W., Dynamic observations of the tensile deformation and fracture of pearlite, *Acta Metallurgica,* **26,** 1405, 1978

25. Oriani, R.A., *Berischte der Bunsen Gesellschaft,* **76,** 898, 1972

26. Petch, N.J. and Stables, P., Delayed fracture of metals under static load, *Nature,* **169,** 842, 1952

27. Petch, N.J., The lowering of fracture stress due to surface absorption, *Philosophical Magazine,* **1,** 331, 1956

28. Boniszewski, T., Metallurgical aspects of reheat cracking of weldments in ferritic steels, *Welding Journal,* **51,** 29, 1972

29. Billy, J., Johansson, T., Loberg, B. and Easterling, K.E., Stress-relief heat treatment of submerged-arc welded microalloyed steels, *Metals Technology,* 67, Feb. 1980

30. Dolby, R.E. and Saunders, G.G., *A Review of the Problems of Reheat Cracking in Nuclear Vessel Steels,* The Welding Institute, Report No 18/1976/M, Abington, 1976

31. Alberry, P.J., Chew, B. and Jones, W.K.C., Prior austenite grain growth in heat affect zone of a 0.5 Cr–Mo–V steel, *Metals Technology,* 317, June 1977

32. Alberry, P.J. and Jones, W.K.C., Comparison of mechanical properties of 2 Cr–Mo and 0.5 Cr–Mo–V simulated, heat affected zones, *Metals Technology,* 45, Jan. 1977

33. Ashby, M.F. and Easterling, K.E., A first report on diagrams for grain growth in welds, *Acta Metallurgica,* **30,** 1969, 1982

34. Ashby, M.F., *Materials I,* Cambridge University Course Notes, Department of Engineering, 1981

35. Murr, L.E., *Interfacial Phenomena in Metals and Alloys,* Addison-Wesley, 1975

36. Dyson, B.F. and Taplin, D.M.R., Creep damage accumulation, *Grain Boundaries (Conference Proceedings),* p. 23, ed. Edmonds, D.V., Institute of Metallurgists, Series 3(5), April 1976

37. Alexander L. Kielland – *ulykken,* (in Norwegian) Norges offentlige utredning, NOU 1981: 11, University Press, Oslo, 1981

Further reading

Christoffel, S.R.J., Nippes, E.F. and Solomon, H.O. (eds), *Weldments: Physical Metallurgy and Failure Phenomena (Conference Proceedings),* 5th Bolton Landing Conference, General Electric Company, 1978

Coe, F.R., *Welding Steels without Hydrogen Cracking,* The Welding Institute, Abington, 1973

Detection and Measurement of Cracks, The Welding Institute, Abington, 1976

Gray, T.G.F., Spence, J. and North, T.H., *Rational Welding Design,* Newnes-Butterworths, 1975

Guide to the Welding and Weldability of C–Mn Steels and C–Mn Microalloyed Steels, International Institute of Welding, Publications Document IIS/IIW–382–71, 1971

Gurney, T.R. and Maddox, S.J., A re-analysis of fatigue data for welded joints in steel, *Welding Research Institute,* **3,** 1, 1973

Harrison, J.D., The state of the art in crack opening displacement testing and analysis, *Metal Construction,* **12,** 1980

Hirth, J. P., Effects of hydrogen on the properties of iron and steel, *Metall. Transactions,* **11A,** 861, 1980

Knott, J.K., *Fundamentals of Fracture Mechanics,* Butterworths, 1973

Richards, K.G., *Weldability of Steel,* The Welding Institute, Abington, 1972

Richards, K.G., *Brittle Fracture of Welded Structures,* The Welding Institute, 1971

Ruge, J., *Handbuch der Schweisstechnik,* vols 1 and 2, Springer-Verlag, 1980

Welding Handbook, vol 4: *Metals and their Weldability,* American Welding Society, 1973

Appendix: Weld cracking tests and weldability formulae

TABLE A.1 Weld cracking tests (after Baker, R.G., Haddrill, D.M. and Roberts, J.E. Assessment of material weldability, *Cracking and Fracture in Welds*, (*Conference Proceedings*), Japanese Welding Society, Tokyo, **C.5. 1–9**, p.II, 1971)

Potential Problem	*Definitive Tests*	*Check Tests*	*Supplementary Tests*
Solidification cracking	Variestraint test Transvariestraint test	Chemical composition	Murex hot-cracking test
Burning and liquation cracking	HAZ mechanical properties assessed using fracture mechanics tests	Chemical composition (Bend tests)	Simulated HAZ, hot tensile testing
Hydrogen induced cracking	Fracture mechanics tests Implant test Tekken test	Chemical composition	Taper hardness test CTS test Y groove test
Lamellar tearing	No entirely suitable tests available (Short transverse tensile testing) (Automated inclusion counting)	Ultrasonic inspection	'Window' test Cruciform test Cranfield test Crack notched bend test T-joint bending test
Stress relief cracking	Fracture mechanics test	Chemical composition	Ring test
Fracture toughness	Fracture mechanics test static — initiation dynamic — propagation Cross-welded wide-plate test Dynamic tear tests	Charpy V-notch	Initiation–Kihara wide plate test deep notch test Propagation – Esso test, Robertson test, etc.
Static strength	Coventional tensile tests on parent material and welded details High temperature tensile tests on parent material and welded details	Tensile tests – parent material	

TABLE A.2 Examples of formulae for estimating weldability indices of C–Mn and low alloy steels (after Hrivňák, I., The mutual relationship and interdependence of developments in steel metallurgy and welding technology, *Welding in the World,* **16,** 1, 1978)

Problem	*Criterion*	*Formula (wt %)*
Carbon equivalent		$C_{equiv} = C + \frac{Mn}{6} + \frac{Cr + Mo + V}{5} + \frac{Ni + Cu}{15}$
Carbon equivalent for the pearlite-free and reduced-pearlite steels		$C^{D}_{equiv} = C + \frac{Si}{25} + \frac{Mn + Cr}{16} + \frac{Cr}{20} + \frac{Ni}{20} + \frac{Mo}{20} + \frac{V}{15}$
Maximum hardness of the underbead zone		$HV_{max} = 90 + 1050\,C + 47\,Si + 75\,Mn + 30\,Ni + 31\,Cr$
Cracking parameter for low-alloy steels		$P_{CM} = C + \frac{Si}{30} + \frac{Mn}{20} + \frac{Cu}{20} + \frac{Ni}{60} + \frac{Cr}{20} + \frac{Mo}{15} + \frac{V}{15}$
Cracking parameter for the low-alloy steels. Cold cracks	The cracks can occur when $P_W > 0$	$P_W = P_{CM} + \frac{H}{60} + \frac{K}{40 \times 10^3}$: H-glycerine test, $H = 0.64 H_{IIW} - 0.93$ For butt joints, $K = 66\,s$ (s = sheet thickness, mm)
Cracking parameter for cold cracks	The cracks occur when $P_s > 1$, the cracks do not occur when $P_s < 0.5$	$P_s = \log \frac{V_R}{V_1} + \frac{H}{10} + \frac{K}{5000}$ V_R is the actual cooling rate of the HAZ at 300°C; V_1 is the critical rate for martensitic reaction; H is the amount of diffusible hydrogen in the base metal (IIW); K is the intensity of restraint for the butt joints $K = 66s$
		$\log V_1 = 3.00 - (4.62\,C + 1.05\,Mn + 0.54\,Ni + 0.50\,Cr + 0.66\,Mo)$

Cracking parameter for cold cracks	The cracks do not occur when $P_{NB} < 0.25\%$	$P_{NB} = C + \frac{Si}{20} + \frac{Mn}{10} + \frac{Cu}{20} + \frac{Cr}{30} + \frac{Mo}{20}$
Calculation of preheat temperature		$T = 1.440\, P_W - 392$ [°C] $T = 350 \{ \frac{1}{360} [360\,C + 40\,(Mn + Cr) + 20\,Ni + 28\,Mo] \times [1 + 0.005\, s] - 0.25\}^{1/2}$ [°C] s = sheet thickness in mm
Hot cracking susceptibility (HCS)	Cracks do not occur when HCS < 4, but for the low-alloy steels HCS = 1.6 + 2	$HCS = \frac{C(S + P + \frac{Si}{25} + \frac{Ni}{100}) \times 10^3}{3Mn + Cr + Mo + V}$
Susceptibility of steel to stress-relief cracking	$P_{SR} \leqslant 0$	$P_{SR} = Cr + Cu + 2\,Mo + 7\,Nb - 5\,Ti - 2$

Index